STEPHEN WOLFRAM

# TWENTY YEARS OF A NEW KIND OF SCIENCE

# STEPHEN WOLFRAM

# TWENTY YEARS OF A NEW KIND OF SCIENCE

*Twenty Years of A New Kind of Science*

Copyright © 2022 Stephen Wolfram, LLC

Wolfram Media, Inc. | wolfram-media.com

ISBN-978-1-57955-049-3 (hardback)
ISBN-978-1-57955-051-6 (ebook)

Science/Computers

Cataloging-in-publication data available at wolfr.am/NKS20-cip

For permission to reproduce images, contact permissions@wolfram.com.

Typeset with Wolfram Notebooks: wolfram.com/notebooks

Printed by Friesens, Manitoba, Canada. ∞ Acid-free paper. First edition. First printing.

# Contents

Twenty Years Later: The Surprising Greater Implications of *A New Kind of Science*     3

*A New Kind of Science*: A 15-Year View     23

It's Been 10 Years: What's Happened with *A New Kind of Science*?     51

Living a Paradigm Shift: Looking Back on Reactions to *A New Kind of Science*     67

At 10 Years: Looking to the Future of *A New Kind of Science*     77

The Making of *A New Kind of Science*     89

Charting a Course for "Complexity": Metamodeling, Ruliology and More     247

Today We Put a Prize on a Small Turing Machine     273

The Prize Is Won; The Simplest Universal Turing Machine Is Proved     275

Announcing the Rule 30 Prizes     285

Gallery of Art     321

# Twenty Years Later: The Surprising Greater Implications of *A New Kind of Science*

*Published May 16, 2022*

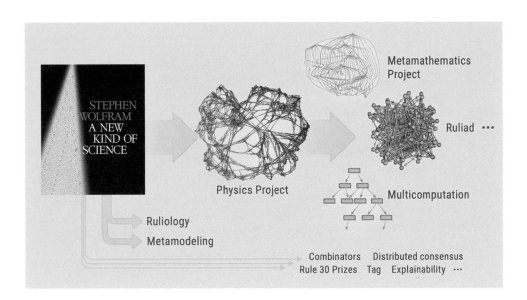

**From the Foundations Laid by *A New Kind of Science***

When *A New Kind of Science* was published twenty years ago I thought what it had to say was important. But what's become increasingly clear—particularly in the last few years—is that it's actually even much more important than I ever imagined. My original goal in *A New Kind of Science* was to take a step beyond the mathematical paradigm that had defined the state of the art in science for three centuries—and to introduce a new paradigm based on computation and on the exploration of the computational universe of possible programs. And already in *A New Kind of Science* one can see that there's immense richness to what can be done with this new paradigm.

There's a new abstract basic science—that I now call ruliology—that's concerned with studying the detailed properties of systems with simple rules. There's a vast new source of "raw material" to "mine" from the computational universe, both for making models of things and for developing technology. And there are new, computational ways to think about fundamental features of how systems in nature and elsewhere work.

But what's now becoming clear is that there's actually something still bigger, still more overarching that the paradigm of *A New Kind of Science* lays the foundations for. In a sense, *A New Kind of Science* defines how one can use computation to think about things. But what we're now realizing is that actually computation is not just a way to think about things: it is at a very fundamental level what everything actually is.

One can see this as a kind of ultimate limit of *A New Kind of Science*. What we call the ruliad is the entangled limit of all possible computations. And what we, for example, experience as physical reality is in effect just our particular sampling of the ruliad. And it's the ideas of *A New Kind of Science*—and particularly things like the Principle of Computational Equivalence—that lay the foundations for understanding how this works.

When I wrote *A New Kind of Science* I discussed the possibility that there might be a way to find a fundamental model of physics based on simple programs. And from that seed has now come the Wolfram Physics Project, which, with its broad connections to existing mathematical physics, now seems to show that, yes, it's really true that our physical universe is "computational all the way down".

But there's more. It's not just that at the lowest level there's some specific rule operating on a vast network of atoms of space. It's that underneath everything is all possible computation, encapsulated in the single unique construct that is the ruliad. And what determines our experience—and the science we use to summarize it—is what characteristics we as observers have in sampling the ruliad.

There is a tower of ideas that relate to fundamental questions about the nature of existence, and the foundations not only of physics, but also of mathematics, computer science and a host of other fields. And these ideas build crucially on the paradigm of *A New Kind of Science*. But they need something else as well: what I now call the multicomputational paradigm. There were hints of it in *A New Kind of Science* when I discussed multiway systems. But it has only been within the past couple of years that this whole new paradigm has begun to come into focus. In *A New Kind of Science* I explored some of the remarkable things that individual computations out in the computational universe can do. What the multicomputational paradigm now does is to consider the aggregate of multiple computations—and in the end the entangled limit of all possible computations, the ruliad.

The Principle of Computational Equivalence is in many ways the intellectual culmination of *A New Kind of Science*—and it has many deep consequences. And one of them is the idea—and uniqueness—of the ruliad. The Principle of Computational Equivalence provides a very general statement about what all possible computational systems do. What the ruliad then does is to pull together the behaviors and relationships of all these systems into a single object that is, in effect, an ultimate representation of everything computational, and indeed in a certain sense simply of everything.

## The Intellectual Journey: From Physics to Physics, and Beyond

The publication of *A New Kind of Science* 20 years ago was for me already the culmination of an intellectual journey that had begun more than 25 years earlier. I had started in theoretical physics as a teenager in the 1970s. And stimulated by my needs in physics, I had then built my first computational language. A couple of years later I returned to basic science, now interested in some very fundamental questions. And from my blend of experience in physics and computing I was led to start trying to formulate things in terms of computation, and computational experiments. And soon discovered the remarkable fact that in the computational universe, even very simple programs can generate immensely complex behavior.

For several years I studied the basic science of the particular class of simple programs known as cellular automata—and the things I saw led me to identify some important general phenomena, most notably computational irreducibility. Then in 1986—having "answered most of the obvious questions I could see"—I left basic science again, and for five years concentrated on creating Mathematica and what's now the Wolfram Language. But in 1991 I took the tools I'd built, and again immersed myself in basic science. The decade that followed brought a long string of exciting and unexpected discoveries about the computational universe and its implications—leading finally in 2002 to the publication of *A New Kind of Science*.

In many ways, *A New Kind of Science* is a very complete book—that in its 1280 pages does well at "answering all the obvious questions", save, notably, for some about the "application area" of fundamental physics. For a couple of years after the book was published, I continued to explore some of these remaining questions. But pretty soon I was swept up in the building of Wolfram|Alpha and then the Wolfram Language, and in all the complicated and often deep questions involved in for the first time creating a full-scale computational language. And so for nearly 17 years I did almost no basic science.

The ideas of *A New Kind of Science* nevertheless continued to exert a deep influence—and I came to see my decades of work on computational language as ultimately being about creating a bridge between the vast capabilities of the computational universe revealed by *A New Kind of Science*, and the specific kinds of ways we humans are able to think about things. This point of view led me to all kinds of important conclusions about the role of computation and its implications for the future. But through all this I kept on thinking that one day I should look at physics again. And finally in 2019, stimulated by a small technical breakthrough, as well as enthusiasm from physicists of a new generation, I decided it was time to try diving into physics again.

My practical tools had developed a lot since I'd worked on *A New Kind of Science*. And—as I have found so often—the passage of years had given me greater clarity and perspective about what I'd discovered in *A New Kind of Science*. And it turned out we were rather quickly able to make spectacular progress. *A New Kind of Science* had introduced definite ideas about how fundamental physics might work. Now we could see that these ideas were very much on the right track, but on their own they did not go far enough. Something else was needed.

In *A New Kind of Science* I'd introduced what I called multiway systems, but I'd treated them as a kind of sideshow. Now—particularly tipped off by quantum mechanics—we realized that multiway systems were not a sideshow but were actually in a sense the main event. They had come out of the computational paradigm of *A New Kind of Science*, but they were really harbingers of a new paradigm: the multicomputational paradigm.

In *A New Kind of Science*, I'd already talked about space—and everything else in the universe—ultimately being made up of a network of discrete elements that I'd now call "atoms of space". And I'd talked about time being associated with the inexorable progressive application of computationally irreducible rules. But now we were thinking not just of a single thread of computation, but instead of a whole multiway system of branching and merging threads—representing in effect a multicomputational history for the universe.

In *A New Kind of Science* I'd devoted a whole chapter to "Processes of Perception and Analysis", recognizing the importance of the observer in computational systems. But with multicomputation there was yet more focus on this, and on how a physical observer knits things together to form a coherent thread of experience. Indeed, it became clear that it's certain features of the observer that ultimately determine the laws of physics we perceive. And in particular it seems that as soon as we—somehow reflecting core features of our conscious experience—believe that we exist persistently through time, but are computationally bounded, then it follows that we will attribute to the universe the central known laws of spacetime and quantum mechanics.

At the level of atoms of space and individual threads of history everything is full of computational irreducibility. But the key point is that observers like us don't experience this; instead we sample certain computationally reducible features—that we can describe in terms of meaningful "laws of physics".

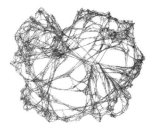

I never expected it would be so easy, but by early 2020—only a few months into our Wolfram Physics Project—we seemed to have successfully identified how the "machine code" of our universe must work. *A New Kind of Science* had established that computation was a powerful way of thinking about things. But now it was becoming clear that actually our whole universe is in a sense "computational all the way down".

But where did this leave the traditional mathematical view? To my surprise, far from being at odds it seemed as if our computation-all-the-way-down model of physics perfectly plugged into a great many of the more abstract existing mathematical approaches. Mediated by multicomputation, the concepts of *A New Kind of Science*—which began as an effort to go beyond mathematics—seemed now to be finding a kind of ultimate convergence with mathematics.

But despite our success in working out the structure of the "machine code" for our universe, a major mystery remained. Let's say we could find a particular rule that could generate everything in our universe. Then we'd have to ask "Why this rule, and not another?" And if "our rule" was simple, how come we'd "lucked out" like that? Ever since I was working on *A New Kind of Science* I'd wondered about this.

And just as we were getting ready to announce the Physics Project in May 2020 the answer began to emerge. It came out of the multicomputational paradigm. And in a sense it was an ultimate version of it. Instead of imagining that the universe follows some particular rule—albeit applying it multicomputationally in all possible ways—what if the universe follows all possible rules?

And then we realized: this is something much more general than physics. And in a sense it's the ultimate computational construct. It's what one gets if one takes all the programs in the computational universe that I studied in *A New Kind of Science* and runs them together—as a single, giant, multicomputational system. It's a single, unique object that I call the ruliad, formed as the entangled limit of all possible computations.

There's no choice about the ruliad. Everything about it is abstractly necessary—emerging as it does just from the formal concept of computation. *A New Kind of Science* developed the abstraction of thinking about things in terms of computation. The ruliad takes this to its ultimate limit—capturing the whole entangled structure of all possible computations—and defining an object that in some sense describes everything.

Once we believe—as the Principle of Computational Equivalence implies—that things like our universe are computational, it then inevitably follows that they are described by the ruliad. But the observer has a crucial role here. Because while as a matter of theoretical science we can discuss the whole ruliad, our experience of it inevitably has to be based on sampling it according to our actual capabilities of perception.

In the end, it's deeply analogous to something that—as I mention in *A New Kind of Science*—first got me interested in fundamental questions in science 50 years ago: the Second Law of thermodynamics. The molecules in a gas move around and interact according to certain rules. But as *A New Kind of Science* argues, one can think about this as a computational process, which can show computational irreducibility. If one didn't worry about the "mechanics" of the observer, one might imagine that one could readily "see through" this computational irreducibility, to the detailed behavior of the molecules underneath. But the point is that a realistic, computationally bounded observer—like us—will be forced by computational irreducibility to perceive only certain "coarse-grained" aspects of what's going on, and so will consider the gas to be behaving in a standard large-scale thermodynamic way.

And so it is, at a grander level, with the ruliad. Observers like us can only perceive certain aspects of what's going on in the ruliad, and a key result of our Physics Project is that with only quite loose constraints on what we're like as observers, it's inevitable that we will perceive our universe to operate according to particular precise known laws of physics. And indeed the attributes that we associate with "consciousness" seem closely tied to what's needed to get the features of spacetime and quantum mechanics that we know from physics. In *A New Kind of Science* one of the conclusions is that the Principle of Computational Equivalence implies a fundamental equivalence between systems (like us) that we consider "intelligent" or "conscious", and systems that we consider "merely computational".

But what's now become clear in the multicomputational paradigm is that there's more to this story. It's not (as people have often assumed) that there's something more powerful about "conscious observers" like us. Actually, it's rather the opposite: that in order to have consistent "conscious experience" we have to have certain limitations (in particular, computational boundedness, and a belief of persistence in time), and these limitations are what make us "see the ruliad" in the way that corresponds to our usual view of the physical world.

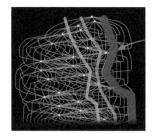

The concept of the ruliad is a powerful one, with implications that significantly transcend the traditional boundaries of science. For example, last year I realized that thinking in terms of the ruliad potentially provides a meaningful answer to the ultimate question of why our universe exists. The answer, I posit, is that the ruliad—as a "purely formal" object—"necessarily exists". And what we perceive as "our universe" is then just the "slice" that corresponds to what we can "see" from the particular place in "rulial space" at which we happen to be. There has to be "something there"—and the remarkable fact is that for an observer with our general characteristics, that something has to have features that are like our usual laws of physics.

In *A New Kind of Science* I discussed how the Principle of Computational Equivalence implies that almost any system can be thought of as being "like a mind" (as in, "the weather has a mind of its own"). But the issue—that for example is of central importance in talking about extraterrestrial intelligence—is how similar to us that mind is. And now with the ruliad we have a more definite way to discuss this. Different minds (even different human ones) can be thought of as being at different places in the ruliad, and thus in effect attributing different rules to the universe. The Principle of Computational Equivalence implies that there must ultimately be a way to translate (or, in effect, move) from one place to another. But the question is how far it is.

Our senses and measuring devices—together with our general paradigms for thinking about things—define the basic area over which our understanding extends, and for which we can readily produce a high-level narrative description of what's going on. And in the past we might have assumed that this was all we'd ever need to reach with whatever science we built. But what *A New Kind of Science*—and now the ruliad—show us is that there's much more out there. There's a whole computational universe of possible programs—many of which behave in ways that are far from our current domain of high-level understanding.

Traditional science we can view as operating by gradually expanding our domain of understanding. But in a sense the key methodological idea that launched *A New Kind of Science* is to do computational experiments, which in effect just "jump without prior understanding" out into the wilds of the computational universe. And that's in the end why all that ruliology in *A New Kind of Science* at first looks so alien: we've effectively jumped quite far from our familiar place in rulial space, so there's no reason to expect we'll recognize anything. And in effect, as the title of the book says, we need to be doing a new kind of science.

In *A New Kind of Science*, an important part of the story has to do with the phenomenon of computational irreducibility, and the way in which it prevents any computationally bounded observer (like us) from being able to "reduce" the behavior of systems, and thereby perceive them as anything other than complex. But now that we're thinking not just about computation, but about multicomputation, other attributes of other observers start to be important too. And with the ruliad ultimately representing everything, the question of what will be perceived in any particular case devolves into one about the characteristics of observers.

In *A New Kind of Science* I give examples of how the same kinds of simple programs (such as cellular automata) can provide good "metamodels" for a variety of kinds of systems in nature and elsewhere, that show up in very different areas of science. But one feature of different areas of science is that they're often concerned with different kinds of questions. And with the focus on the characteristics of the observer this is something we get to capture—and we get to discuss, for example, what the chemical observer, or the economic observer, might be like, and how that affects their perception of what's ultimately in the ruliad.

In Chapter 12 of *A New Kind of Science* there's a long section on "Implications for Mathematics and Its Foundations", which begins with the observation that just as many models in science seem to be able to start from simple rules, mathematics is traditionally specifically set up to start from simple axioms. I then analyzed how multiway systems could be thought of as defining possible derivations (or proofs) of new mathematical theorems from axioms or other theorems—and I discussed how the difficulty of doing mathematics can be thought of as a reflection of computational irreducibility.

But informed by our Physics Project I realized that there's much more to say about the foundations of mathematics—and this has led to our recently launched Metamathematics Project. At the core of this project is the idea that mathematics, like physics, is ultimately just a sampling of the ruliad. And just as the ruliad defines the lowest-level machine code of physics, so does it also for mathematics.

The traditional axiomatic level of mathematics (with its built-in notions of variables and operators and so on) is already higher level than the "raw ruliad". And a crucial observation is that just like physical observers operate at a level far above things like the atoms of space, so "mathematical observers" mostly operate at a level far above the raw ruliad, or even the "assembly code" of axioms. In an analogy with gases, the ruliad—or even axiom systems—are talking about the "molecular dynamics" level; but "mathematical observers" operate more at the "fluid dynamics" level.

And the result of this is what I call the physicalization of metamathematics: the realization that our "perception" of mathematics is like our perception of physics. And that, for example, the very possibility of consistently doing higher-level mathematics where we don't always have to drop down to the level of axioms or the raw ruliad has the same origin as the fact that "observers like us" typically view space as something continuous, rather than something made up of lots of atoms of space.

In *A New Kind of Science* I considered it a mystery why phenomena like undecidability are not more common in typical pure mathematics. But now our Metamathematics Project provides an answer that's based on the character of mathematical observers.

My stated goal at the beginning of *A New Kind of Science* was to go beyond the mathematical paradigm, and that's exactly what was achieved. But now there's almost a full circle—because we see that building on *A New Kind of Science* and the computational paradigm we reach the multicomputational paradigm and the ruliad, and then we realize that mathematics, like physics, is part of the ruliad. Or, put another way, mathematics, like physics—and like everything else—is "made of computation", and all computation is in the ruliad.

And that means that insofar as we consider there to be physical reality, so also we must consider there to be "mathematical reality". Physical reality arises from the sampling of the ruliad by physical observers; so similarly mathematical reality must arise from the sampling of the ruliad by mathematical observers. Or, in other words, if we believe that the physical world exists, so we must—essentially like Plato—also believe that the mathematics exists, and that there is an underlying reality to mathematics.

All of these ideas rest on what was achieved in *A New Kind of Science* but now go significantly beyond it. In an "Epilog" that I eventually cut from the final version of *A New Kind of Science* I speculated that "major new directions" might be built in 15–30 years. And when I wrote that, I wasn't really expecting that I would be the one to be central in doing that. And indeed I suspect that had I simply continued the direct path in basic science defined by my work on *A New Kind of Science*, it wouldn't have been me.

It's not something I've explicitly planned, but at this point I can look back on my life so far and see it as a repeated alternation between technology and basic science. Each builds on the other, giving me both ideas and tools—and creating in the end a taller and taller intellectual tower. But what's crucial is that every alternation is in many ways a fresh start, where I'm able to use what I've done before, but have a chance to reexamine everything from a new perspective. And so it has been in the past few years with *A New Kind of Science*: having returned to basic science after 17 years away, it's been possible to make remarkably rapid and dramatic progress that's taken things to a new and wholly unexpected level.

## The Arrival of a Fourth Scientific Paradigm

In the course of intellectual history, there've been very few fundamentally different paradigms introduced for theoretical science. The first is what one might call the "structural paradigm", in which one's basically just concerned with what things are made of. And beginning in antiquity—and continuing for two millennia—this was pretty much the only paradigm on offer. But in the 1600s there was, as I described it in the opening sentence of *A New Kind of Science*, a "dramatic new idea"—that one could describe not just how things are, but also what they can do, in terms of mathematical equations.

And for three centuries this "mathematical paradigm" defined the state of the art for theoretical science. But as I went on to explain in the opening paragraph of *A New Kind of Science*, my goal was to develop a new "computational paradigm" that would describe things not in terms of mathematical equations but instead in terms of computational rules or programs. There'd been precursors to this in my own work in the 1980s, but despite the practical use of computers in applying the mathematical paradigm, there wasn't much of a concept of describing things, say in nature, in a fundamentally computational way.

One feature of a mathematical equation is that it aims to encapsulate "in one fell swoop" the whole behavior of a system. Solve the equation and you'll know everything about what the system will do. But in the computational paradigm it's a different story. The underlying computational rules for a system in principle determine what it will do. But to actually find out what it does, you have to run those rules—which is often a computationally irreducible process.

Put another way: in the structural paradigm, one doesn't talk about time at all. In the mathematical paradigm, time is there, but it's basically just a parameter, that if you can solve the equations you can set to whatever value you want. In the computational paradigm, however, time is something more fundamental: it's associated with the actual irreducible progression of computation in a system.

It's an important distinction that cuts to the core of theoretical science. Heavily influenced by the mathematical paradigm, it's often been assumed that science is fundamentally about being able to make predictions, or in a sense having a model that can "outrun" the system you're studying, and say what it's going to do with much less computational effort than the system itself.

But computational irreducibility implies that there's a fundamental limit to this. There are systems whose behavior is in effect "too complex" for us to ever be able to "find a formula for it". And this is not something we could, for example, resolve just by increasing our mathematical sophistication: it is a fundamental limit that arises from the whole structure of the computational paradigm. In effect, from deep inside science we're learning that there are fundamental limitations on what science can achieve.

But as I mentioned in *A New Kind of Science*, computational irreducibility has an upside as well. If everything were computationally reducible, the passage of time wouldn't in any fundamental sense add up to anything; we'd always be able to "jump ahead" and see what the outcome of anything would be without going through the steps, and we'd never have something we could reasonably experience as free will.

In practical computing it's pretty common to want to go straight from "question" to "answer", and not be interested in "what happened inside". But in *A New Kind of Science* there is in a sense an immediate emphasis on "what happens inside". I don't just show the initial input and final output for a cellular automaton. I show its whole "spacetime" history. And now that we have a computational theory of fundamental physics we can see that all the richness of our physical experience is contained in the "process inside". We don't just want to know the endpoint of the universe; we want to live the ongoing computational process that corresponds to our experience of the passage of time.

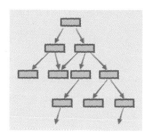

But, OK, so in *A New Kind of Science* we reached what we might identify as the third major paradigm for theoretical science. But the exciting—and surprising—thing is that inspired by our Physics Project we can now see a fourth paradigm: the multicomputational paradigm. And while the computational paradigm involves considering the progression of particular computations, the multicomputational paradigm involves considering the entangled progression of many computations. The computational paradigm involves a single thread of time. The multicomputational paradigm involves multiple threads of time that branch and merge.

What in a sense forced us into the multicomputational paradigm was thinking about quantum mechanics in our Physics Project, and realizing that multicomputation was inevitable in our models. But the idea of multicomputation is vastly more general, and in fact immediately applies to any system where at any given step multiple things can happen. In *A New Kind of Science* I studied many kinds of computational systems—like cellular automata and Turing machines—where one definite thing happens at each step. I looked a little at multiway systems—primarily ones based on string rewriting. But now in general in the multicomputational paradigm one is interested in studying multiway systems of all kinds. They can be based on simple iterations, say involving numbers, in which multiple functions can be applied at each step. They can be based on systems like games where there are multiple moves at each step. And they can be based on a whole range of systems in nature, technology and elsewhere where there are multiple "asynchronous" choices of events that can occur.

Given the basic description of multicomputational systems, one might at first assume that whatever difficulties there are in deducing the behavior of computational systems, they would only be greater for multicomputational systems. But the crucial point is that whereas with a purely computational system (like a cellular automaton) it's perfectly reasonable to imagine "experiencing" its whole evolution—say just by seeing a picture of it, the same is not true of a multicomputational system. Because for observers like us, who fundamentally experience time in a single thread, we have no choice but to somehow "sample" or "coarse grain" a multicomputational system if we are to reduce its behavior to something we can "experience".

And there's then a remarkable formal fact: if one has a system that shows fundamental computational irreducibility, then computationally bounded "single-thread-of-time" observers inevitably perceive certain effective behavior in the system, that follows something like the typical laws of physics. Once again we can make an analogy with gases made from large numbers of molecules. Large-scale (computationally bounded) observers will essentially inevitably perceive gases to follow, say, the standard gas laws, quite independent of the detailed properties of individual molecules.

In other words, the interplay between an "observer like us" and a multicomputational system will effectively select out a slice of computational reducibility from the underlying computational irreducibility. And although I didn't see this coming, it's in the end fairly obvious that something like this has to happen. The Principle of Computational Equivalence makes it basically inevitable that the underlying processes in the universe will be computationally irreducible. But somehow the particular features of the universe that we perceive and care about have to be ones that have enough computational reducibility that we can, for example, make consistent decisions about what to do, and we're not just continually confronted by irreducible unpredictability.

So how general can we expect this picture of multicomputation to be, with its connection to the kinds of things we've seen in physics? It seems to be extremely general, and to provide a true fourth paradigm for theoretical science.

There are many kinds of systems for which the multicomputational paradigm seems to be immediately relevant. Beyond physics and metamathematics, there seems to be near-term promise in chemistry, molecular biology, evolutionary biology, neuroscience, immunology, linguistics, economics, machine learning, distributed computing and more. In each case there are underlying low-level elements (such as molecules) that interact through some kind of events (say collisions or reactions). And then there's a big question of what the relevant observer is like.

In chemistry, for example, the observer could just measure the overall concentration of some kind of molecule, coarse-graining together all the individual instances of those molecules. Or the observer could be sensitive, for example, to detailed causal relationships between collisions among molecules. In traditional chemistry, things like this generally aren't "observed". But in biology (for example in connection with membranes), or in molecular computing, they may be crucial.

When I began the project that became *A New Kind of Science* the central question I wanted to answer is why we see so much complexity in so many kinds of systems. And with the computational paradigm and the ubiquity of computational irreducibility we had an answer, which also in a sense told us why it was difficult to make certain kinds of progress in a whole range of areas.

But now we've got a new paradigm, the multicomputational paradigm. And the big surprise is that through the intermediation of the observer we can tap into computational reducibility, and potentially find "physics-like" laws for all sorts of fields. This may not work for the questions that have traditionally been asked in these fields. But the point is that with the "right kind of observer" there's computational reducibility to be found. And that computational reducibility may be something we can tap into for understanding, or to use some kind of system for technology.

It can all be seen as starting with the ruliad, and involving almost philosophical questions of what one can call "observer theory". But in the end it gives us very practical ideas and methods that I think have the potential to lead to unexpectedly dramatic progress in a remarkable range of fields.

I knew that *A New Kind of Science* would have practical applications, particularly in modeling, in technology and in producing creative material. And indeed it has. But for our Physics Project applications seemed much further away, perhaps centuries. But a great surprise has been that through the multicomputational paradigm it seems as if there are going to be some quite immediate and very practical applications of the Physics Project.

In a sense the reason for this is that through the intermediation of multicomputation we see that many kinds of systems share the same underlying "metastructure". And this means that as soon as there are things to say about one kind of system these can be applied to other systems. And in particular the great successes of physics can be applied to a whole range of systems that share the same multicomputational metastructure.

An immediate example is in practical computing, and particularly in the Wolfram Language. It's something of a personal irony that the Wolfram Language is based on transformation rules for symbolic expressions, which is a structure very similar to what ends up being what's involved in the Physics Project. But there's a crucial difference: in the usual case of the Wolfram Language, everything works in a purely computational way, with a particular transformation being done at each step. But now there's the potential to generalize that to the multicomputational case, and in effect to trace the multiway system of every possible transformation.

It's not easy to pick out of that structure things that we can readily understand. But there are important lessons from physics for this. And as we build out the multicomputational capabilities of the Wolfram Language I fully expect that the "notational clarity" it will bring will help us to formulate much more in terms of the multicomputational paradigm.

I built the Wolfram Language as a tool that would help me explore the computational paradigm, and from that paradigm there emerged principles like the Principle of Computational Equivalence, which in turn led me to see the possibility of something like Wolfram|Alpha. But now from the latest basic science built on the foundations of *A New Kind of Science*, together with the practical tooling of the Wolfram Language, it's becoming possible again to see how to make conceptual advances that can drive technology that will again in turn let us make—likely dramatic—progress in basic science.

### Harvesting Seeds from *A New Kind of Science*

*A New Kind of Science* is full of intellectual seeds. And in the past few years—having now returned to basic science—I've been harvesting a few of those seeds. The Physics Project and the Metamathematics Project are two major results. But there's been quite a bit more. And in fact it's rather remarkable how many things that were barely more than footnotes in *A New Kind of Science* have turned into major projects, with important results.

Back in 2018—a year before beginning the Physics Project—I returned, for example, to what's become known as the Wolfram Axiom: the axiom that I found in *A New Kind of Science* that is the very simplest possible axiom for Boolean algebra. But my focus now was not so much on the axiom itself as on the automated process of proving its correctness, and the effort to see the relation between "pure computation" and what one might consider a human-absorbable "narrative proof".

Computational irreducibility appeared many times, notably in my efforts to understand AI ethics and the implications of computational contracts. I've no doubt that in the years to come, the concept of computational irreducibility will become increasingly important in everyday thinking—a bit like how concepts such as energy and momentum from the mathematical paradigm have become important. And in 2019, for example, computational irreducibility made an appearance in government affairs, as a result of me testifying about its implications for legislation about AI selection of content on the internet.

In *A New Kind of Science* I explored many specific systems about which one can ask all sorts of questions. And one might think that after 20 years "all the obvious questions" would have been answered. But they have not. And in a sense the fact that they have not is a direct reflection of the ubiquity of computational irreducibility. But it's a fundamental feature that whenever there's computational irreducibility, there must also be pockets of computational reducibility: in other words, the very existence of computational irreducibility implies an infinite frontier of potential progress.

Back in 2007, we'd had great success with our Turing Machine Prize, and the Turing machine that I'd suspected was the very simplest possible universal Turing machine was indeed proved universal—providing another piece of evidence for the Principle of Computational Equivalence. And in a sense there's a general question that's raised by *A New Kind of Science* about where the threshold of universality—or computational equivalence—really is in different kinds of systems.

But there are simpler-to-define questions as well. And ever since I first studied rule 30 in 1984 I'd wondered about many questions related to it. And in October 2019 I decided to launch the Rule 30 Prizes, defining three specific easy-to-state questions about rule 30. So far I don't know of progress on them. And for all I know they'll be open problems for centuries. From the point of view of the ruliad we can think of them as distant explorations in rulial space, and the question of when they can be answered is like the question of when we'll have the technology to get to some distant place in physical space.

Having launched the Physics Project in April 2020, it was rapidly clear that its ideas could also be applied to metamathematics. And it even seemed as if it might be easier to make relevant "real-world" observations in metamathematics than in physics. And the seed for this was in a note in *A New Kind of Science* entitled "Empirical Metamathematics". That note contained one picture of the theorem-dependency graph of Euclid's *Elements*, which in the summer of 2020 expanded into a 70-page study. And in my recent "Physicalization of Metamathematics" there's a continuation of that—beginning to map out empirical metamathematical space, as explored in the practice of mathematics, with the idea that multicomputational phenomena that in physics may take technically infeasible particle accelerators or telescopes might actually be within reach.

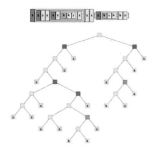

In addition to being the year we launched our Physics Project, 2020 was also the 100th anniversary of combinators—the first concrete formalization of universal computation. In *A New Kind of Science* I devoted a few pages and some notes to combinators, but I decided to do a deep dive and use both what I'd learned from *A New Kind of Science* and from the Physics Project to take a new look at them. Among other things the result was another application of multicomputation, as well as the realization that even though the S, K combinators from 1920 seemed very minimal, it was possible that S alone might also be universal, though with something different than the usual input→output "workflow" of computation.

In *A New Kind of Science* a single footnote mentions multiway Turing machines. And early last year I turned this seed into a long and detailed study that provides further foundational examples of multicomputation, and explores the question of just what it means to "do a computation" multicomputationally—something which I believe is highly relevant not only for practical distributed computing but also for things like molecular computing.

In 2021 it was the centenary of Post tag systems, and again I turned a few pages in *A New Kind of Science* into a long and detailed study. And what's important about both this and my study of combinators is that they provide foundational examples (much like cellular automata in *A New Kind of Science*), which even in the past year or so I've used multiple times in different projects.

In mid-2021, yet another few-page discussion in *A New Kind of Science* turned into a detailed study of "The Problem of Distributed Consensus". And once again, this turned out to have a multicomputational angle, at first in understanding the multiway character of possible outcomes, but later with the realization that the formation of consensus is deeply related to the process of measurement and the coarse-graining involved in it—and the fundamental way that observers extract "coherent experiences" from systems.

In *A New Kind of Science*, there's a short note about multiway systems based on numbers. And once again, in fall 2021 I expanded on this to produce an extensive study of such systems, as a certain kind of very minimal example of multicomputation, that at least in some cases connects with traditional mathematical ideas.

From the vantage point of multicomputation and our Physics Project it's interesting to look back at *A New Kind of Science*, and see some of what it describes with more clarity. In the fall of 2021, for example, I reviewed what had become of the original goal of "understanding complexity", and what methodological ideas had emerged from that effort. I identified two primary ones, which I called "ruliology" and "metamodeling". Ruliology, as I've mentioned above, is my new name for the pure, basic science of studying the behavior of systems with simple rules: in effect, it's the science of exploring the computational universe.

Metamodeling is the key to making connections to systems in nature and elsewhere that one wants to study. Its goal is to find the "minimal models for models". Often there are existing models for systems. But the question is what the ultimate essence of those models is. Can everything be reduced to a cellular automaton? Or a multiway system? What is the minimal "computational essence" of a system? And as we begin to apply the multicomputational paradigm to different fields, a key step will be metamodeling.

Ruliology and metamodeling are in a sense already core concepts in *A New Kind of Science*, though not under those names. Observer theory is much less explicitly covered. And many concepts—like branchial space, token-event graphs, the multiway causal graph and the ruliad—have only emerged now, with the Physics Project and the arrival of the multicomputational paradigm.

Multicomputation, the Physics Project and the Metamathematics Project are sowing their own seeds. But there are still many more seeds to harvest even from *A New Kind of Science*. And just as the multicomputational paradigm was not something that I, for one, could foresee from *A New Kind of Science*, no doubt there will in time be other major new directions that will emerge. But, needless to say, one should expect that it will be computationally irreducible to determine what will happen: a metacontribution of the science to the consideration of its own future.

## The Doing of Science

The creation of *A New Kind of Science* took me a decade of intense work, none of which saw the light of day until the moment the book was published on May 14, 2002. Returning to basic science 17 years later the world had changed and it was possible for me to adopt a quite different approach, in a sense making the process of doing science as open and incremental as possible.

It's helped that there's the web, the cloud and livestreaming. But in a sense the most crucial element has been the Wolfram Language, and its character as a full-scale computational language. Yes, I use English to tell the story of what we're doing. But fundamentally I'm doing science in the Wolfram Language, using it both as a practical tool, and as a medium for organizing my thoughts, and sharing and communicating what I'm doing.

Starting in 2003, we've had an annual Wolfram Summer School at which a long string of talented students have explored ideas based on *A New Kind of Science*, always through the medium of the Wolfram Language. In the last couple of years we've added a Physics track, connected to the Physics Project, and this year we're adding a Metamathematics track, connected to the Metamathematics Project.

During the 17 years that I wasn't focused on basic science, I was doing technology development. And I think it's fair to say that at Wolfram Research over the past 35 years we've created a remarkably effective "machine" for doing innovative research and development. Mostly it's been producing technology and products. But one of the very interesting features of the Physics Project and the projects that have followed it is that we've been applying the same managed approach to innovation to them that we have been using so successfully for so many years at our company. And I consider the results to be quite spectacular: in a matter of weeks or months I think we've managed to deliver what might otherwise have taken years, if it could have been done at all.

And particularly with the arrival of the multicomputational paradigm there's quite a challenge. There are a huge number of exceptionally promising directions to follow, that have the potential to deliver revolutionary results. And with our concepts of managed research, open science and broad connection to talent it should be possible to make great progress even fairly quickly. But to do so requires significant scaling up of our efforts so far, which is why we're now launching the Wolfram Institute to serve as a focal point for these efforts.

When I think about *A New Kind of Science*, I can't help but be struck by all the things that had to align to make it possible. My early experiences in science and technology, the personal environment I'd created—and the tools I built. I wondered at the time whether the five years I took "away from basic science" to launch Mathematica and what's now the Wolfram Language might have slowed down what became *A New Kind of Science*. Looking back I can say that the answer was definitively no. Because without the Wolfram Language the creation of *A New Kind of Science* would have needed "not just a decade", but likely more than a lifetime.

And a similar pattern has repeated now, though even more so. The Physics Project and everything that has developed from it has been made possible by a tower of specific circumstances that stretch back nearly half a century—including my 17-year hiatus from basic science. Had all these circumstances not aligned, it is hard to say when something like the Physics Project would have happened, but my guess is that it would have been at least a significant part of a century away.

It is a lesson of the history of science that the absorption of major new paradigms is a slow process. And normally the timescales are long compared to the 20 years since *A New Kind of Science* was published. But in a sense we've managed to jump far ahead of schedule with the Physics Project and with the development of the multicomputational paradigm. Five years ago, when I summarized the first 15 years of *A New Kind of Science* I had no idea that any of this would happen.

But now that it has—and with all the methodology we've developed for getting science done—it feels as if we have a certain obligation to see just what can be achieved. And to see just what can be built in the years to come on the foundations laid down by *A New Kind of Science*.

# *A New Kind of Science:*
# A 15-Year View

*Published May 16, 2017*

It's now 15 years since I published my book *A New Kind of Science*—more than 25 since I started writing it, and more than 35 since I started working towards it. But with every passing year I feel I understand more about what the book is really about—and why it's important. I wrote the book, as its title suggests, to contribute to the progress of science. But as the years have gone by, I've realized that the core of what's in the book actually goes far beyond science—into many areas that will be increasingly important in defining our whole future.

So, viewed from a distance of 15 years, what is the book really about? At its core, it's about something profoundly abstract: the theory of all possible theories, or the universe of all possible universes. But for me one of the achievements of the book is the realization that one can explore such fundamental things concretely—by doing actual experiments in the computational universe of possible programs. And in the end the book is full of what might at first seem like quite alien pictures made just by running very simple such programs.

Back in 1980, when I made my living as a theoretical physicist, if you'd asked me what I thought simple programs would do, I expect I would have said "not much". I had been very interested in the kind of complexity one sees in nature, but I thought—like a typical reductionistic scientist—that the key to understanding it must lie in figuring out detailed features of the underlying component parts.

In retrospect I consider it incredibly lucky that all those years ago I happened to have the right interests and the right skills to actually try what is in a sense the most basic experiment in the computational universe: to systematically take a sequence of the simplest possible programs, and run them.

I could tell as soon as I did this that there were interesting things going on, but it took a couple more years before I began to really appreciate the force of what I'd seen. For me it all started with one picture:

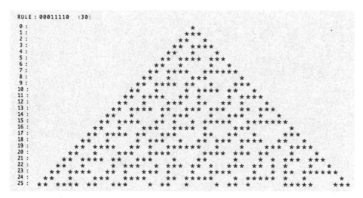

Or, in modern form:

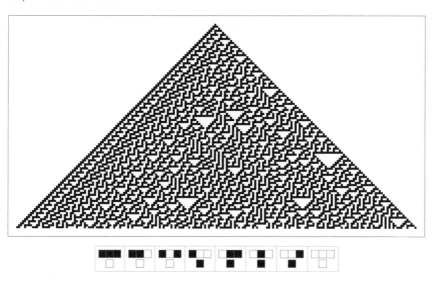

I call it rule 30. It's my all-time favorite discovery, and today I carry it around everywhere on my business cards. What is it? It's one of the simplest programs one can imagine. It operates on rows of black and white cells, starting from a single black cell, and then repeatedly applies the rules at the bottom. And the crucial point is that even though those rules are by any measure extremely simple, the pattern that emerges is not.

It's a crucial—and utterly unexpected—feature of the computational universe: that even among the very simplest programs, it's easy to get immensely complex behavior. It took me a solid decade to understand just how broad this phenomenon is. It doesn't just happen in programs ("cellular automata") like rule 30. It basically shows up whenever you start enumerating possible rules or possible programs whose behavior isn't obviously trivial.

Similar phenomena had actually been seen for centuries in things like the digits of $\pi$ and the distribution of primes—but they were basically just viewed as curiosities, and not as signs of something profoundly important. It's been nearly 35 years since I first saw what happens in rule 30, and with every passing year I feel I come to understand more clearly and deeply what its significance is.

Four centuries ago it was the discovery of the moons of Jupiter and their regularities that sowed the seeds for modern exact science, and for the modern scientific approach to thinking. Could my little rule 30 now be the seed for another such intellectual revolution, and a new way of thinking about everything?

In some ways I might personally prefer not to take responsibility for shepherding such ideas ("paradigm shifts" are hard and thankless work). And certainly for years I have just quietly used such ideas to develop technology and my own thinking. But as computation and AI become increasingly central to our world, I think it's important that the implications of what's out there in the computational universe be more widely understood.

## Implications of the Computational Universe

Here's the way I see it today. From observing the moons of Jupiter we came away with the idea that—if looked at right—the universe is an ordered and regular place, that we can ultimately understand. But now, in exploring the computational universe, we quickly come upon things like rule 30 where even the simplest rules seem to lead to irreducibly complex behavior.

One of the big ideas of *A New Kind of Science* is what I call the Principle of Computational Equivalence. The first step is to think of every process—whether it's happening with black and white squares, or in physics, or inside our brains—as a computation that somehow transforms input to output. What the Principle of Computational Equivalence says is that above an extremely low threshold, all processes correspond to computations of equivalent sophistication.

It might not be true. It might be that something like rule 30 corresponds to a fundamentally simpler computation than the fluid dynamics of a hurricane, or the processes in my brain as I write this. But what the Principle of Computational Equivalence says is that in fact all these things are computationally equivalent.

It's a very important statement, with many deep implications. For one thing, it implies what I call computational irreducibility. If something like rule 30 is doing a computation just as sophisticated as our brains or our mathematics, then there's no way we can "outrun" it: to figure out what it will do, we have to do an irreducible amount of computation, effectively tracing each of its steps.

The mathematical tradition in exact science has emphasized the idea of predicting the behavior of systems by doing things like solving mathematical equations. But what computational irreducibility implies is that out in the computational universe that often won't work, and instead the only way forward is just to explicitly run a computation to simulate the behavior of the system.

## A Shift in Looking at the World

One of the things I did in *A New Kind of Science* was to show how simple programs can serve as models for the essential features of all sorts of physical, biological and other systems. Back when the book appeared, some people were skeptical about this. And indeed at that time there was a 300-year unbroken tradition that serious models in science should be based on mathematical equations.

But in the past 15 years something remarkable has happened. For now, when new models are created—whether of animal patterns or web browsing behavior—they are overwhelmingly more often based on programs than on mathematical equations.

Year by year, it's been a slow, almost silent, process. But by this point, it's a dramatic shift. Three centuries ago pure philosophical reasoning was supplanted by mathematical equations. Now in these few short years, equations have been largely supplanted by programs. For now, it's mostly been something practical and pragmatic: the models work better, and are more useful.

But when it comes to understanding the foundations of what's going on, one's led not to things like mathematical theorems and calculus, but instead to ideas like the Principle of Computational Equivalence. Traditional mathematics-based ways of thinking have made concepts like force and momentum ubiquitous in the way we talk about the world. But now as we think in fundamentally computational terms we have to start talking in terms of concepts like undecidability and computational irreducibility.

Will some type of tumor always stop growing in some particular model? It might be undecidable. Is there a way to work out how a weather system will develop? It might be computationally irreducible.

These concepts are pretty important when it comes to understanding not only what can and cannot be modeled, but also what can and cannot be controlled in the world. Computational irreducibility in economics is going to limit what can be globally controlled. Computational irreducibility in biology is going to limit how generally effective therapies can be—and make highly personalized medicine a fundamental necessity.

And through ideas like the Principle of Computational Equivalence we can start to discuss just what it is that allows nature—seemingly so effortlessly—to generate so much that seems so complex to us. Or how even deterministic underlying rules can lead to computationally irreducible behavior that for all practical purposes can seem to show "free will".

## Mining the Computational Universe

A central lesson of *A New Kind of Science* is that there's a lot of incredible richness out there in the computational universe. And one reason that's important is that it means that there's a lot of incredible stuff out there for us to "mine" and harness for our purposes.

Want to automatically make an interesting custom piece of art? Just start looking at simple programs and automatically pick out one you like—as in our WolframTones music site from a decade ago. Want to find an optimal algorithm for something? Just search enough programs out there, and you'll find one.

We've normally been used to creating things by building them up, step by step, with human effort—progressively creating architectural plans, or engineering drawings, or lines of code. But the discovery that there's so much richness so easily accessible in the computational universe suggests a different approach: don't try building anything; just define what you want, and then search for it in the computational universe.

Sometimes it's really easy to find. Like let's say you want to generate apparent randomness. Well, then just enumerate cellular automata (as I did in 1984), and very quickly you come upon rule 30—which turns out to be one of the very best known generators of apparent randomness (look down the center column of cell values, for examples). In other situations you might have to search 100,000 cases (as I did in finding the simplest axiom system for logic, or the simplest universal Turing machine), or you might have to search millions or even trillions of cases. But in the past 25 years, we've had incredible success in just discovering algorithms out there in the computational universe—and we rely on many of them in implementing the Wolfram Language.

At some level it's quite sobering. One finds some tiny program out in the computational universe. One can tell it does what one wants. But when one looks at what it's doing, one doesn't have any real idea how it works. Maybe one can analyze some part—and be struck by how "clever" it is. But there just isn't a way for us to understand the whole thing; it's not something familiar from our usual patterns of thinking.

Of course, we've often had similar experiences before—when we use things from nature. We may notice that some particular substance is a useful drug or a great chemical catalyst, but we may have no idea why. But in doing engineering and in most of our modern efforts to build technology, the great emphasis has instead been on constructing things whose design and operation we can readily understand.

In the past we might have thought that was enough. But what our explorations of the computational universe show is that it's not: selecting only things whose operation we can readily understand misses most of the immense power and richness that's out there in the computational universe.

## A World of Discovered Technology

What will the world look like when more of what we have is mined from the computational universe? Today the environment we build for ourselves is dominated by things like simple shapes and repetitive processes. But the more we use what's out there in the computational universe, the less regular things will look. Sometimes they may look a bit "organic", or like what we see in nature (since after all, nature follows similar kinds of rules). But sometimes they may look quite random, until perhaps suddenly and incomprehensibly they achieve something we recognize.

For several millennia we as a civilization have been on a path to understand more about what happens in our world—whether by using science to decode nature, or by creating our own environment through technology. But to use more of the richness of the computational universe we must at least to some extent forsake this path.

In the past, we somehow counted on the idea that between our brains and the tools we could create we would always have fundamentally greater computational power than the things around us—and as a result we would always be able to "understand" them. But what the Principle of Computational Equivalence says is that this isn't true: out in the computational universe there are lots of things just as powerful as our brains or the tools we build. And as soon as we start using those things, we lose the "edge" we thought we had.

Today we still imagine we can identify discrete "bugs" in programs. But most of what's powerful out there in the computational universe is rife with computational irreducibility—so the only real way to see what it does is just to run it and watch what happens.

We ourselves, as biological systems, are a great example of computation happening at a molecular scale—and we are no doubt rife with computational irreducibility (which is, at some fundamental level, why medicine is hard). I suppose it's a tradeoff: we could limit our technology to consist only of things whose operation we understand. But then we would miss all that richness that's out there in the computational universe. And we wouldn't even be able to match the achievements of our own biology in the technology we create.

## Machine Learning and the Neural Net Renaissance

There's a common pattern I've noticed with intellectual fields. They go for decades and perhaps centuries with only incremental growth, and then suddenly, usually as a result of a methodological advance, there's a burst of "hypergrowth" for perhaps 5 years, in which important new results arrive almost every week.

I was fortunate enough that my own very first field—particle physics—was in its period of hypergrowth right when I was involved in the late 1970s. And for myself, the 1990s felt like a kind of personal period of hypergrowth for what became *A New Kind of Science*—and indeed that's why I couldn't pull myself away from it for more than a decade.

But today, the obvious field in hypergrowth is machine learning, or, more specifically, neural nets. It's funny for me to see this. I actually worked on neural nets back in 1981, before I started on cellular automata, and several years before I found rule 30. But I never managed to get neural nets to do anything very interesting—and actually I found them too messy and complicated for the fundamental questions I was concerned with.

And so I "simplified them"—and wound up with cellular automata. (I was also inspired by things like the Ising model in statistical physics, etc.) At the outset, I thought I might have simplified too far, and that my little cellular automata would never do anything interesting. But then I found things like rule 30. And I've been trying to understand its implications ever since.

In building Mathematica and the Wolfram Language, I'd always kept track of neural nets, and occasionally we'd use them in some small way for some algorithm or another. But about 5 years ago I suddenly started hearing amazing things: that somehow the idea of training neural nets to do sophisticated things was actually working. At first I wasn't sure. But then we started building neural net capabilities in the Wolfram Language, and finally two years ago we released our ImageIdentify.com website—and now we've got our whole symbolic neural net system. And, yes, I'm impressed. There are lots of tasks that had traditionally been viewed as the unique domain of humans, but which now we can routinely do by computer.

But what's actually going on in a neural net? It's not really to do with the brain; that was just the inspiration (though in reality the brain probably works more or less the same way). A neural net is really a sequence of functions that operate on arrays of numbers, with each function typically taking quite a few inputs from around the array. It's not so different from a cellular automaton. Except that in a cellular automaton, one's usually dealing with, say, just 0s and 1s, not arbitrary numbers like 0.735.

And instead of taking inputs from all over the place, in a cellular automaton each step takes inputs only from a very well-defined local region.

Now, to be fair, it's pretty common to study "convolutional neural nets", in which the patterns of inputs are very regular, just like in a cellular automaton. And it's becoming clear that having precise (say 32-bit) numbers isn't critical to the operation of neural nets; one can probably make do with just a few bits.

But a big feature of neural nets is that we know how to make them "learn". In particular, they have enough features from traditional mathematics (like involving continuous numbers) that techniques like calculus can be applied to provide strategies to make them incrementally change their parameters to "fit their behavior" to whatever training examples they're given.

It's far from obvious how much computational effort, or how many training examples, will be needed. But the breakthrough of about five years ago was the discovery that for many important practical problems, what's available with modern GPUs and modern web-collected training sets can be enough.

Pretty much nobody ends up explicitly setting or "engineering" the parameters in a neural net. Instead, what happens is that they're found automatically. But unlike with simple programs like cellular automata, where one's typically enumerating all possibilities, in current neural nets there's an incremental process, essentially based on calculus, that manages to progressively improve the net—a little like the way biological evolution progressively improves the "fitness" of an organism.

It's plenty remarkable what comes out from training a neural net in this way, and it's plenty difficult to understand how the neural net does what it does. But in some sense the neural net isn't venturing too far across the computational universe: it's always basically keeping the same basic computational structure, and just changing its behavior by changing parameters.

But to me the success of today's neural nets is a spectacular endorsement of the power of the computational universe, and another validation of the ideas of *A New Kind of Science*. Because it shows that out in the computational universe, away from the constraints of explicitly building systems whose detailed behavior one can foresee, there are immediately all sorts of rich and useful things to be found.

## NKS Meets Modern Machine Learning

Is there a way to bring the full power of the computational universe—and the ideas of *A New Kind of Science*—to the kinds of things one does with neural nets? I suspect so. And in fact, as the details become clear, I wouldn't be surprised if exploration of the computational universe saw its own period of hypergrowth: a "mining boom" of perhaps unprecedented proportions.

In current work on neural nets, there's a definite tradeoff one sees. The more what's going on inside the neural net is like a simple mathematical function with essentially arithmetic parameters, the easier it is to use ideas from calculus to train the network. But the more what's going is like a discrete program, or like a computation whose whole structure can change, the more difficult it is to train the network.

It's worth remembering, though, that the networks we're routinely training now would have looked utterly impractical to train only a few years ago. It's effectively just all those quadrillions of GPU operations that we can throw at the problem that makes training feasible. And I won't be surprised if even quite pedestrian (say, local exhaustive search) techniques will fairly soon let one do significant training even in cases where no incremental numerical approach is possible. And perhaps even it will be possible to invent some major generalization of things like calculus that will operate in the full computational universe. (I have some suspicions, based on thinking about generalizing basic notions of geometry to cover things like cellular automaton rule spaces.)

What would this let one do? Likely it would let one find considerably simpler systems that could achieve particular computational goals. And maybe that would bring within reach some qualitatively new level of operations, perhaps beyond what we're used to being possible with things like brains.

There's a funny thing that's going on with modeling these days. As neural nets become more successful, one begins to wonder: why bother to simulate what's going on inside a system when one can just make a black-box model of its output using a neural net? Well, if we manage to get machine learning to reach deeper into the computational universe, we won't have as much of this tradeoff any more—because we'll be able to learn models of the mechanism as well as the output.

I'm pretty sure that bringing the full computational universe into the purview of machine learning will have spectacular consequences. But it's worth realizing that computational universality—and the Principle of Computational Equivalence—make it less a matter of principle. Because they imply that even neural nets of the kinds we have now are universal, and are capable of emulating anything any other system can do. (In fact, this universality result was essentially what launched the whole modern idea of neural nets, back in 1943.)

And as a practical matter, the fact that current neural net primitives are being built into hardware and so on will make them a desirable foundation for actual technology systems, though, even if they're far from optimal. But my guess is that there are tasks where for the foreseeable future access to the full computational universe will be necessary to make them even vaguely practical.

## Finding AI

What will it take to make artificial intelligence? As a kid, I was very interested in figuring out how to make a computer know things, and be able to answer questions from what it knew. And when I studied neural nets in 1981, it was partly in the context of trying to understand how to build such a system. As it happens, I had just developed SMP, which was a forerunner of Mathematica (and ultimately the Wolfram Language)—and which was very much based on symbolic pattern matching ("if you see this, transform it to that"). At the time, though, I imagined that artificial intelligence was somehow a "higher level of computation", and I didn't know how to achieve it.

I returned to the problem every so often, and kept putting it off. But then when I was working on *A New Kind of Science* it struck me: if I'm to take the Principle of Computational Equivalence seriously, then there can't be any fundamentally "higher level of computation"—so AI must be achievable just with the standard ideas of computation that I already know.

And it was this realization that got me started building Wolfram|Alpha. And, yes, what I found is that lots of those very "AI-oriented things", like natural language understanding, could be done just with "ordinary computation", without any magic new AI invention. Now, to be fair, part of what was happening was that we were using ideas and methods from *A New Kind of Science*: we weren't just engineering everything; we were often searching the computational universe for rules and algorithms to use.

So what about "general AI"? Well, I think at this point that with the tools and understanding we have, we're in a good position to automate essentially anything we can define. But definition is a more difficult and central issue than we might imagine.

The way I see things at this point is that there's a lot of computation even near at hand in the computational universe. And it's powerful computation. As powerful as anything that happens in our brains. But we don't recognize it as "intelligence" unless it's aligned with our human goals and purposes.

Ever since I was writing *A New Kind of Science*, I've been fond of quoting the aphorism "the weather has a mind of its own". It sounds so animistic and pre-scientific. But what the Principle of Computational Equivalence says is that actually, according to the most modern science, it's true: the fluid dynamics of the weather is the same in its computational sophistication as the electrical processes that go on in our brains.

But is it "intelligent"? When I talk to people about *A New Kind of Science*, and about AI, I'll often get asked when I think we'll achieve "consciousness" in a machine. Life, intelligence, consciousness: they are all concepts that we have a specific example of, here on Earth. But what are they in general? All life on Earth shares RNA and the structure of cell membranes. But surely that's just because all life we know is part of one connected thread of history; it's not that such details are fundamental to the very concept of life.

And so it is with intelligence. We have only one example we're sure of: us humans. (We're not even sure about animals.) But human intelligence as we experience it is deeply entangled with human civilization, human culture and ultimately also human physiology—even though none of these details are presumably relevant in the abstract definition of intelligence.

We might think about extraterrestrial intelligence. But what the Principle of Computational Equivalence implies is that actually there's "alien intelligence" all around us. But somehow it's just not quite aligned with human intelligence. We might look at rule 30, for example, and be able to see that it's doing sophisticated computation, just like our brains. But somehow it just doesn't seem to have any "point" to what it's doing.

We imagine that in doing the things we humans do, we operate with certain goals or purposes. But rule 30, for example, just seems to be doing what it's doing—just following some definite rule. In the end, though, one realizes we're not so very different. After all, there are definite laws of nature that govern our brains. So anything we do is at some level just playing out those laws.

Any process can actually be described either in terms of mechanism ("the stone is moving according to Newton's laws"), or in terms of goals ("the stone is moving so as to minimize potential energy"). The description in terms of mechanism is usually what's most useful in connecting with science. But the description in terms of goals is usually what's most useful in connecting with human intelligence.

And this is crucial in thinking about AI. We know we can have computational systems whose operations are as sophisticated as anything. But can we get them to do things that are aligned with human goals and purposes?

In a sense this is what I now view as the key problem of AI: it's not about achieving underlying computational sophistication, but instead it's about communicating what we want from this computation.

## The Importance of Language

I've spent much of my life as a computer language designer—most importantly creating what is now the Wolfram Language. I'd always seen my role as a language designer being to imagine the possible computations people might want to do, then—like a reductionist scientist—trying to "drill down" to find good primitives from which all these computations could be built up. But somehow from *A New Kind of Science*, and from thinking about AI, I've come to think about it a little differently.

Now what I more see myself as doing is making a bridge between our patterns of human thinking, and what the computational universe is capable of. There are all sorts of amazing things that can in principle be done by computation. But what the language does is to provide a way for us humans to express what we want done, or want to achieve—and then to get this actually executed, as automatically as possible.

Language design has to start from what we know and are familiar with. In the Wolfram Language, we name the built-in primitives with English words, leveraging the meanings that those words have acquired. But the Wolfram Language is not like natural language. It's something more structured, and more powerful. It's based on the words and concepts that we're familiar with through the shared corpus of human knowledge. But it gives us a way to build up arbitrarily sophisticated programs that in effect express arbitrarily complex goals.

Yes, the computational universe is capable of remarkable things. But they're not necessarily things that we humans can describe or relate to. But in building the Wolfram Language my goal is to do the best I can in capturing everything we humans want—and being able to express it in executable computational terms.

When we look at the computational universe, it's hard not to be struck by the limitations of what we know how to describe or think about. Modern neural nets provide an interesting example. For the ImageIdentify function of the Wolfram Language we've trained a neural net to identify thousands of kinds of things in the world. And to cater to our human purposes, what the network ultimately does is to describe what it sees in terms of concepts that we can name with words—tables, chairs, elephants, etc.

But internally what the network is doing is to identify a series of features of any object in the world. Is it green? Is it round? And so on. And what happens as the neural network is trained is that it identifies features it finds useful for distinguishing different kinds of things in the world. But the point is that almost none of these features are ones to which we happen to have assigned words in human language.

Out in the computational universe it's possible to find what may be incredibly useful ways to describe things. But they're alien to us humans. They're not something we know how to express, based on the corpus of knowledge our civilization has developed.

Now of course new concepts are being added to the corpus of human knowledge all the time. Back a century ago, if someone saw a nested pattern they wouldn't have any way to describe it. But now we'd just say "it's a fractal". But the problem is that in the computational universe there's an infinite collection of "potentially useful concepts"—with which we can never hope to ultimately keep up.

## The Analogy in Mathematics

When I wrote *A New Kind of Science* I viewed it in no small part as an effort to break away from the use of mathematics—at least as a foundation for science. But one of the things I realized is that the ideas in the book also have a lot of implications for pure mathematics itself.

What is mathematics? Well, it's a study of certain abstract kinds of systems, based on things like numbers and geometry. In a sense it's exploring a small corner of the computational universe of all possible abstract systems. But still, plenty has been done in mathematics: indeed, the 3 million or so published theorems of mathematics represent perhaps the largest single coherent intellectual structure that our species has built.

Ever since Euclid, people have at least notionally imagined that mathematics starts from certain axioms (say, $a+b=a+b$, $a+0=a$, etc.), then builds up derivations of theorems. Why is math hard? The answer is fundamentally rooted in the phenomenon of computational irreducibility—which here is manifest in the fact that there's no general way to shortcut the series of steps needed to derive a theorem. In other words, it can be arbitrarily hard to get a result in mathematics. But worse than that—as Gödel's theorem showed—there can be mathematical statements where there just aren't any finite ways to prove or disprove them from the axioms. And in such cases, the statements just have to be considered "undecidable".

And in a sense what's remarkable about math is that one can usefully do it at all. Because it could be that most mathematical results one cares about would be undecidable. So why doesn't that happen?

Well, if one considers arbitrary abstract systems it happens a lot. Take a typical cellular automaton—or a Turing machine—and ask whether it's true that the system, say, always settles down to periodic behavior regardless of its initial state. Even something as simple as that will often be undecidable.

So why doesn't this happen in mathematics? Maybe there's something special about the particular axioms used in mathematics. And certainly if one thinks they're the ones that uniquely describe science and the world there might be a reason for that. But one of the whole points of the book is that actually there's a whole computational universe of possible rules that can be useful for doing science and describing the world.

And in fact I don't think there's anything abstractly special about the particular axioms that have traditionally been used in mathematics: I think they're just accidents of history.

What about the theorems that people investigate in mathematics? Again, I think there's a strong historical character to them. For all but the most trivial areas of mathematics, there's a whole sea of undecidability out there. But somehow mathematics picks the islands where theorems can actually be proved—often particularly priding itself on places close to the sea of undecidability where the proof can only be done with great effort.

I've been interested in the whole network of published theorems in mathematics (it's a thing to curate, like wars in history, or properties of chemicals). And one of the things I'm curious about is whether there's an inexorable sequence to the mathematics that's done, or whether, in a sense, random parts are being picked.

And here, I think, there's a considerable analogy to the kind of thing we were discussing before with language. What is a proof? Basically it's a way of explaining to someone why something is true. I've made all sorts of automated proofs in which there are hundreds of steps, each perfectly verifiable by computer. But—like the innards of a neural net—what's going on looks alien and not understandable by a human.

For a human to understand, there have to be familiar "conceptual waypoints". It's pretty much like with words in languages. If some particular part of a proof has a name ("Smith's theorem"), and has a known meaning, then it's useful to us. But if it's just a lump of undifferentiated computation, it won't be meaningful to us.

In pretty much any axiom system, there's an infinite set of possible theorems. But which ones are "interesting"? That's really a human question. And basically it's going to end up being ones with "stories". In the book I show that for the simple case of basic logic, the theorems that have historically been considered interesting enough to be given names happen to be precisely the ones that are in some sense minimal.

But my guess is that for richer axiom systems pretty much anything that's going to be considered "interesting" is going to have to be reached from things that are already considered interesting. It's like building up words or concepts: you don't get to introduce new ones unless you can directly relate them to existing ones.

In recent years I've wondered quite a bit about how inexorable or not progress is in a field like mathematics. Is there just one historical path that can be taken, say from arithmetic to algebra to the higher reaches of modern mathematics? Or are there an infinite diversity of possible paths, with completely different histories for mathematics?

The answer is going to depend on—in a sense—the "structure of metamathematical space": just what is the network of true theorems that avoid the sea of undecidability? Maybe it'll be different for different fields of mathematics, and some will be more "inexorable" (so it feels like the math is being "discovered") than others (where it seems more like the math is arbitrary, and "invented").

But to me one of the most interesting things is how close—when viewed in these kinds of terms—questions about the nature and character of mathematics end up being to questions about the nature and character of intelligence and AI. And it's this kind of commonality that makes me realize just how powerful and general the ideas in *A New Kind of Science* actually are.

## When Is There a Science?

There are some areas of science—like physics and astronomy—where the traditional mathematical approach has done quite well. But there are others—like biology, social science and linguistics—where it's had a lot less to say. And one of the things I've long believed is that what's needed to make progress in these areas is to generalize the kinds of models one's using, to consider a broader range of what's out there in the computational universe.

And indeed in the past 15 or so years there's been increasing success in doing this. And there are lots of biological and social systems, for example, where models have now been constructed using simple programs.

But unlike with mathematical models which can potentially be "solved", these computational models often show computational irreducibility, and are typically used by doing explicit simulations. This can be perfectly successful for making particular predictions, or for applying the models in technology. But a bit like for the automated proofs of mathematical theorems one might still ask, "is this really science?".

Yes, one can simulate what a system does, but does one "understand" it? Well, the problem is that computational irreducibility implies that in some fundamental sense one can't always "understand" things. There might be no useful "story" that can be told; there may be no "conceptual waypoints"—only lots of detailed computation.

Imagine that one's trying to make a science of how the brain understands language—one of the big goals of linguistics. Well, perhaps we'll get an adequate model of the precise rules which determine the firing of neurons or some other low-level representation of the brain. And then we look at the patterns generated in understanding some whole collection of sentences.

Well, what if those patterns look like the behavior of rule 30? Or, closer at hand, the innards of some recurrent neural network? Can we "tell a story" about what's happening? To do so would basically require that we create some kind of higher-level symbolic representation: something where we effectively have words for core elements of what's going on.

But computational irreducibility implies that there may ultimately be no way to create such a thing. Yes, it will always be possible to find patches of computational reducibility, where some things can be said. But there won't be a complete story that can be told. And one might say there won't be a useful reductionistic piece of science to be done. But that's just one of the things that happens when one's dealing with (as the title says) a new kind of science.

## Controlling the AIs

People have gotten very worried about AI in recent years. They wonder what's going to happen when AIs "get much smarter" than us humans. Well, the Principle of Computational Equivalence has one piece of good news: at some fundamental level, AIs will never be "smarter"—they'll just be able to do computations that are ultimately equivalent to what our brains do, or, for that matter, what all sorts of simple programs do.

As a practical matter, of course, AIs will be able to process larger amounts of data more quickly than actual brains. And no doubt we'll choose to have them run many aspects of the world for us—from medical devices, to central banks to transportation systems, and much more.

So then it's important to figure how we'll tell them what to do. As soon as we're making serious use of what's out there in the computational universe, we're not going to be able to give a line-by-line description of what the AIs are going to do. Rather, we're going to have to define goals for the AIs, then let them figure out how best to achieve those goals.

In a sense we've already been doing something like this for years in the Wolfram Language. There's some high-level function that describes something you want to do ("lay out a graph", "classify data", etc.). Then it's up to the language to automatically figure out the best way to do it.

And in the end the real challenge is to find a way to describe goals. Yes, you want to search for cellular automata that will make a "nice carpet pattern", or a "good edge detector". But what exactly do those things mean? What you need is a language that a human can use to say as precisely as possible what they mean.

It's really the same problem as I've been talking about a lot here. One has to have a way for humans to be able to talk about things they care about. There's infinite detail out there in the computational universe. But through our civilization and our shared cultural history we've come to identify certain concepts that are important to us. And when we describe our goals, it's in terms of these concepts.

Three hundred years ago people like Leibniz were interested in finding a precise symbolic way to represent the content of human thoughts and human discourse. He was far too early. But now I think we're finally in a position to actually make this work. In fact, we've already gotten a long way with the Wolfram Language in being able to describe real things in the world. And I'm hoping it'll be possible to construct a fairly complete "symbolic discourse language" that lets us talk about the things we care about.

Right now we write legal contracts in "legalese" as a way to make them slightly more precise than ordinary natural language. But with a symbolic discourse language we'll be able to write true "computational contracts" that describe in high-level terms what we want to have happen—and then machines will automatically be able to verify or execute the contract.

But what about the AIs? Well, we need to tell them what we generally want them to do. We need to have a contract with them. Or maybe we need to have a constitution for them. And it'll be written in some kind of symbolic discourse language, that both allows us humans to express what we want, and is executable by the AIs.

There's lots to say about what should be in an AI Constitution, and how the construction of such things might map onto the political and cultural landscape of the world. But one of the obvious questions is: can the constitution be simple, like Asimov's Laws of Robotics?

And here what we know from *A New Kind of Science* tells us the answer: it can't be. In a sense the constitution is an attempt to sculpt what can happen in the world and what can't. But computational irreducibility says that there will be an unbounded collection of cases to consider.

For me it's interesting to see how theoretical ideas like computational irreducibility end up impinging on these very practical—and central—societal issues. Yes, it all started with questions about things like the theory of all possible theories. But in the end it turns into issues that everyone in society is going to end up being concerned about.

### There's an Endless Frontier

Will we reach the end of science? Will we—or our AIs—eventually invent everything there is to be invented?

For mathematics, it's easy to see that there's an infinite number of possible theorems one can construct. For science, there's an infinite number of possible detailed questions to ask. And there's also an infinite array of possible inventions one can construct.

But the real question is: will there always be interesting new things out there? Well, computational irreducibility says there will always be new things that need an irreducible amount of computational work to reach from what's already there. So in a sense there'll always be "surprises", that aren't immediately evident from what's come before.

But will it just be like an endless array of different weirdly shaped rocks? Or will there be fundamental new features that appear, that we humans consider interesting?

It's back to the very same issue we've encountered several times before: for us humans to find things "interesting" we have to have a conceptual framework that we can use to think about them. Yes, we can identify a "persistent structure" in a cellular automaton. Then maybe we can start talking about "collisions between structures". But when we just see a whole mess of stuff going on, it's not going to be "interesting" to us unless we have some higher-level symbolic way to talk about it.

In a sense, then, the rate of "interesting discovery" isn't going to be limited by our ability to go out into the computational universe and find things. Instead, it's going to be limited by our ability as humans to build a conceptual framework for what we're finding.

It's a bit like what happened in the whole development of what became *A New Kind of Science*. People had seen related phenomena for centuries if not millennia (distribution of primes, digits of π, etc.). But without a conceptual framework they just didn't seem "interesting", and nothing was built around them. And indeed as I understand more about what's out there in the computational universe—and even about things I saw long ago there—I gradually build up a conceptual framework that lets me go further.

By the way, it's worth realizing that inventions work a little differently from discoveries. One can see something new happen in the computational universe, and that might be a discovery. But an invention is about figuring out how something can be achieved in the computational universe.

And—like in patent law—it isn't really an invention if you just say "look, this does that". You have to somehow understand a purpose that it's achieving.

In the past, the focus of the process of invention has tended to be on actually getting something to work ("find the lightbulb filament that works", etc.). But in the computational universe, the focus shifts to the question of what you want the invention to do. Because once you've described the goal, finding a way to achieve it is something that can be automated.

That's not to say that it will always be easy. In fact, computational irreducibility implies that it can be arbitrarily difficult. Let's say you know the precise rules by which some chemicals can interact. Can you find a chemical synthesis pathway that will let you get to some particular chemical structure? There may be a way, but computational irreducibility implies that there may be no way to find out how long the pathway may be. And if you haven't found a pathway you may never be sure if it's because there isn't one, or just because you didn't reach it yet.

## The Fundamental Theory of Physics

If one thinks about reaching the edge of science, one cannot help but wonder about the fundamental theory of physics. Given everything we've seen in the computational universe, is it conceivable that our physical universe could just correspond to one of those programs out there in the computational universe?

Of course, we won't really know until or unless we find it. But in the years since *A New Kind of Science* appeared, I've become ever more optimistic about the possibilities. Needless to say, it would be a big change for physics. Today there are basically two major frameworks for thinking about fundamental physics: general relativity and quantum field theory. General relativity is a bit more than 100 years old; quantum field theory maybe 90. And both have achieved spectacular things. But neither has succeeded in delivering us a complete fundamental theory of physics. And if nothing else, I think after all this time, it's worth trying something new.

But there's another thing: from actually exploring the computational universe, we have a huge amount of new intuition about what's possible, even in very simple models. We might have thought that the kind of richness we know exists in physics would require some very elaborate underlying model. But what's become clear is that that kind of richness can perfectly well emerge even from a very simple underlying model.

What might the underlying model be like? I'm not going to discuss this in great detail here, but suffice it to say that I think the most important thing about the model is that it should have as little as possible built in. We shouldn't have the hubris to think we know how the universe is constructed; we should just take a general type of model that's as unstructured as possible, and do what we typically do in the computational universe: just search for a program that does what we want.

My favorite formulation for a model that's as unstructured as possible is a network: just a collection of nodes with connections between them. It's perfectly possible to formulate such a model as an algebraic-like structure, and probably many other kinds of things. But we can think of it as a network. And in the way I've imagined setting it up, it's a network that's somehow "underneath" space and time: every aspect of space and time as we know it must emerge from the actual behavior of the network.

Over the past decade or so there's been increasing interest in things like loop quantum gravity and spin networks. They're related to what I've been doing in the same way that they also involve networks. And maybe there's some deeper relationship. But in their usual formulation, they're much more mathematically elaborate.

From the point of view of the traditional methods of physics, this might seem like a good idea. But with the intuition we have from studying the computational universe—and using it for science and technology—it seems completely unnecessary. Yes, we don't yet know the fundamental theory of physics. But it seems sensible to start with the simplest hypothesis. And that's definitely something like a simple network of the kind I've studied.

At the outset, it'll look pretty alien to people (including myself) trained in traditional theoretical physics. But some of what emerges isn't so alien. A big result I found nearly 20 years ago (that still hasn't been widely understood) is that when you look at a large enough network of the kind I studied you can show that its averaged behavior follows

Einstein's equations for gravity. In other words, without putting any fancy physics into the underlying model, it ends up automatically emerging. I think it's pretty exciting.

People ask a lot about quantum mechanics. Yes, my underlying model doesn't build in quantum mechanics (just as it doesn't build in general relativity). Now, it's a little difficult to pin down exactly what the essence of "being quantum mechanical" actually is. But there are some very suggestive signs that my simple networks actually end up showing what amounts to quantum behavior—just like in the physics we know.

OK, so how should one set about actually finding the fundamental theory of physics if it's out there in the computational universe of possible programs? Well, the obvious thing is to just start searching for it, starting with the simplest programs.

I've been doing this—more sporadically than I would like—for the past 15 years or so. And my main discovery so far is that it's actually quite easy to find programs that aren't obviously not our universe. There are plenty of programs where space or time are obviously completely different from the way they are in our universe, or there's some other pathology. But it turns out it's not so difficult to find candidate universes that aren't obviously not our universe.

But we're immediately bitten by computational irreducibility. We can simulate the candidate universe for billions of steps. But we don't know what it's going to do—and whether it's going to grow up to be like our universe, or completely different.

It's pretty unlikely that in looking at that tiny fragment of the very beginning of a universe we're going to ever be able to see anything familiar, like a photon. And it's not at all obvious that we'll be able to construct any kind of descriptive theory, or effective physics. But in a sense the problem is bizarrely similar to the one we have even in systems like neural networks: there's computation going on there, but can we identify "conceptual waypoints" from which we can build up a theory that we might understand?

It's not at all clear our universe has to be understandable at that level, and it's quite possible that for a very long time we'll be left in the strange situation of thinking we might have "found our universe" out in the computational universe, but not being sure.

Of course, we might be lucky, and it might be possible to deduce an effective physics, and see that some little program that we found ends up reproducing our whole universe. It would be a remarkable moment for science. But it would immediately raise a host of new questions—like why this universe, and not another?

## Box of a Trillion Souls

Right now us humans exist as biological systems. But in the future it's certainly going to be technologically possible to reproduce all the processes in our brains in some purely digital—computational—form. So insofar as those processes represent "us", we're going to be able to be "virtualized" on pretty much any computational substrate. And in this case we might imagine that the whole future of a civilization could wind up in effect as a "box of a trillion souls".

Inside that box there would be all kinds of computations going on, representing the thoughts and experiences of all those disembodied souls. Those computations would reflect the rich history of our civilization, and all the things that have happened to us. But at some level they wouldn't be anything special.

It's perhaps a bit disappointing, but the Principle of Computational Equivalence tells us that ultimately these computations will be no more sophisticated than the ones that go on in all sorts of other systems—even ones with simple rules, and no elaborate history of civilization. Yes, the details will reflect all that history. But in a sense without knowing what to look for—or what to care about—one won't be able to tell that there's anything special about it.

OK, but what about for the "souls" themselves? Will one be able to understand their behavior by seeing that they achieve certain purposes? Well, in our current biological existence, we have all sorts of constraints and features that give us goals and purposes. But in a virtualized "uploaded" form, most of these just go away.

I've thought quite a bit about how "human" purposes might evolve in such a situation, recognizing, of course, that in virtualized form there's little difference between human and AI. The disappointing vision is that perhaps the future of our civilization consists in disembodied souls in effect "playing videogames" for the rest of eternity.

But what I've slowly realized is that it's actually quite unrealistic to project our view of goals and purposes from our experience today into that future situation. Imagine talking to someone from a thousand years ago and trying to explain that people in the future would be walking on treadmills every day, or continually sending photographs to their friends. The point is that such activities don't make sense until the cultural framework around them has developed. It's the same story yet again as with trying to characterize what's interesting or what's explainable. It relies on the development of a whole network of conceptual waypoints.

Can we imagine what the mathematics of 100 years from now will be like? It depends on concepts we don't yet know. So similarly if we try to imagine human motivation in the future, it's going to rely on concepts we don't know. Our best description from today's viewpoint might be that those disembodied souls are just "playing videogames". But to them there might be a whole subtle motivation structure that they could only explain by rewinding all sorts of steps in history and cultural development.

By the way, if we know the fundamental theory of physics then in a sense we can make the virtualization complete, at least in principle: we can just run a simulation of the universe for those disembodied souls. Of course, if that's what's happening, then there's no particular reason it has to be a simulation of our particular universe. It could as well be any universe from out in the computational universe.

Now, as I've mentioned, even in any given universe one will never in a sense run out of things to do, or discover. But I suppose I myself at least find it amusing to imagine that at some point those disembodied souls might get bored with just being in a simulated version of our physical universe—and might decide it's more fun (whatever that means to them) to go out and explore the broader computational universe. Which would mean that in a sense the future of humanity would be an infinite voyage of discovery in the context of none other than *A New Kind of Science*!

## The Economics of the Computational Universe

Long before we have to think about disembodied human souls, we'll have to confront the issue of what humans should be doing in a world where more and more can be done automatically by AIs. Now in a sense this issue is nothing new: it's just an extension of the long-running story of technology and automation. But somehow this time it feels different.

And I think the reason is in a sense just that there's so much out there in the computational universe, that's so easy to get to. Yes, we can build a machine that automates some particular task. We can even have a general-purpose computer that can be programmed to do a full range of different tasks. But even though these kinds of automation extend what we can do, it still feels like there's effort that we have to put into them.

But the picture now is different—because in effect what we're saying is that if we can just define the goal we want to achieve, then everything else will be automatic. All sorts of computation, and, yes, "thinking", may have to be done, but the idea is that it's just going to happen, without human effort.

At first, something seems wrong. How could we get all that benefit, without putting in more effort? It's a bit like asking how nature could manage to make all the complexity it does—even though when we build artifacts, even with great effort, they end up far less complex. The answer, I think, is it's mining the computational universe. And it's exactly the same thing for us: by mining the computational universe, we can achieve essentially an unbounded level of automation.

If we look at the important resources in today's world, many of them still depend on actual materials. And often these materials are literally mined from the Earth. Of course, there are accidents of geography and geology that determine by whom and where that mining can be done. And in the end there's a limit (if often very large) to the amount of material that'll ever be available.

But when it comes to the computational universe, there's in a sense an inexhaustible supply of material—and it's accessible to anyone. Yes, there are technical issues about how to "do the mining", and there's a whole stack of technology associated with doing it well. But the ultimate resource of the computational universe is a global and infinite one. There's no scarcity and no reason to be "expensive". One just has to understand that it's there, and take advantage of it.

**The Path to Computational Thinking**

Probably the greatest intellectual shift of the past century has been the one towards the computational way of thinking about things. I've often said that if one picks almost any field "X", from archaeology to zoology, then by now there either is, or soon will be, a field called "computational X"—and it's going to be the future of the field.

I myself have been deeply involved in trying to enable such computational fields, in particular through the development of the Wolfram Language. But I've also been interested in what is essentially the meta problem: how should one teach abstract computational thinking, for example to kids? The Wolfram Language is certainly important as a practical tool. But what about the conceptual, theoretical foundations?

Well, that's where *A New Kind of Science* comes in. Because at its core it's discussing the pure abstract phenomenon of computation, independent of its applications to particular fields or tasks. It's a bit like with elementary mathematics: there are things to teach and understand just to introduce the ideas of mathematical thinking, independent of their specific applications. And so it is too with the core of

*A New Kind of Science*. There are things to learn about the computational universe that give intuition and introduce patterns of computational thinking—quite independent of detailed applications.

One can think of it as a kind of "pre-computer science", or "pre-computational X". Before one gets into discussing the specifics of particular computational processes, one can just study the simple but pure things one finds in the computational universe.

And, yes, even before kids learn to do arithmetic, it's perfectly possible for them to fill out something like a cellular automaton coloring book—or to execute for themselves or on a computer a whole range of different simple programs. What does it teach? Well, it certainly teaches the idea that there can be definite rules or algorithms for things—and that if one follows them one can create useful and interesting results. And, yes, it helps that systems like cellular automata make obvious visual patterns, that for example one can even find in nature (say on mollusc shells).

As the world becomes more computational—and more things are done by AIs and by mining the computational universe—there's going to be an extremely high value not only in understanding computational thinking, but also in having the kind of intuition that develops from exploring the computational universe and that is, in a sense, the foundation for *A New Kind of Science*.

## What's Left to Figure Out?

My goal over the decade that I spent writing *A New Kind of Science* was, as much as possible, to answer all the first round of "obvious questions" about the computational universe. And looking back 15 years later I think that worked out pretty well. Indeed, today, when I wonder about something to do with the computational universe, I find it's incredibly likely that somewhere in the main text or notes of the book I already said something about it.

But one of the biggest things that's changed over the past 15 years is that I've gradually begun to understand more of the implications of what the book describes. There are lots of specific ideas and discoveries in the book. But in the longer term I think what's most significant is how they serve as foundations, both practical and conceptual, for a whole range of new things that one can now understand and explore.

But even in terms of the basic science of the computational universe, there are certainly specific results one would still like to get. For example, it would be great to get more evidence for or against the Principle of Computational Equivalence, and its domain of applicability.

Like most general principles in science, the whole epistemological status of the Principle of Computational Equivalence is somewhat complicated. Is it like a mathematical theorem that can be proved? Is it like a law of nature that might (or might not) be true about the universe? Or is it like a definition, say of the very concept of computation? Well, much like, say, the second law of thermodynamics or evolution by natural selection, it's a combination of these.

But one thing that's significant is that it's possible to get concrete evidence for (or against) the Principle of Computational Equivalence. The principle says that even systems with very simple rules should be capable of arbitrarily sophisticated computation—so that in particular they should be able to act as universal computers.

And indeed one of the results of the book is that this is true for one of the simplest possible cellular automata (rule 110). Five years after the book was published I decided to put up a prize for evidence about another case: the simplest conceivably universal Turing machine. And I was very pleased that in just a few months the prize was won, the Turing machine was proved universal, and there was another piece of evidence for the Principle of Computational Equivalence.

There's a lot to do in developing the applications of *A New Kind of Science*. There are models to be made of all sorts of systems. There's technology to be found. Art to be created. There's also a lot to do in understanding the implications.

But it's important not to forget the pure investigation of the computational universe. In the analogy of mathematics, there are applications to be pursued. But there's also a "pure mathematics" that's worth pursuing in its own right. And so it is with the computational universe: there's a huge amount to explore just at an abstract level. And indeed (as the title of the book implies) there's enough to define a whole new kind of science: a pure science of the computational universe. And it's the opening of that new kind of science that I think is the core achievement of *A New Kind of Science*—and the one of which I am most proud.

# It's Been 10 Years: What's Happened with *A New Kind of Science*?

*Published May 7, 2012*

On May 14, 2012, it'll be 10 years since *A New Kind of Science* ("the NKS book") was published. After 20 years of research, and nearly 11 years writing the book, I'd taken most things about as far as I could at that time. And so when the book was finished, I mainly launched myself back into technology development. And inspired by my work on the NKS book, I'm happy to say that I've had a very fruitful decade (Mathematica reinvented, CDF, Wolfram|Alpha, etc.).

I've been doing little bits of NKS-oriented science here and there (notably at our annual Summer School). But mostly I've been busy with other things. And so it's been other people who've been having the fun of moving the science of NKS forward. But almost every day I'll hear about something that's been being done with NKS. And as we approach the 10-year mark, I've been very curious to try to get at least a slightly more systematic view of what's been going on.

A place to start is the academic literature, where there's now an average of slightly over one new paper per day published citing the NKS book—with that number steadily increasing. The papers span all kinds of areas (here identified by journal fields).

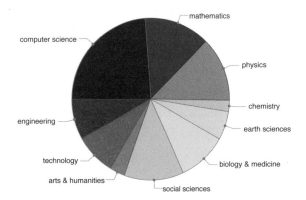

And looking through the list of papers my main response is "Wow—so much stuff". Some of the early papers seem a bit questionable, but as the decade has gone on, my impression from skimming through papers is that people are really "getting it"— understanding the ideas in the NKS book, and making good and interesting use of them.

There are typically three broad categories of NKS work: pure NKS, applied NKS, and the NKS way of thinking.

Pure NKS is about studying the computational universe as basic science for its own sake—investigating simple programs like cellular automata, seeing what they do, and gradually abstracting general principles. Applied NKS is about taking what one finds in the computational universe, and using it as raw material to create models, technology and other things. And the NKS way of thinking is about taking ideas and principles from NKS—like computational irreducibility or the Principle of Computational Equivalence—and using them as a conceptual framework for thinking about things.

And with these categories, here's how the academic papers published in different types of journals break down:

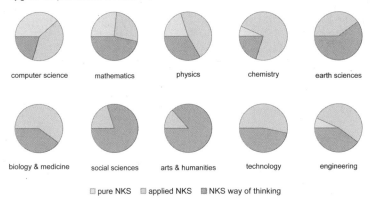

So what are all these papers actually about? Let's start with the largest group: applied NKS. And among these, a striking feature is the development of models for a dizzying array of systems and phenomena. In traditional science, new models are fairly rare. But in just a decade of applied NKS academic literature, there are already hundreds of new models.

Hair patterns in mice. Shapes of human molars. Collective butterfly motion. Evolution of soil thicknesses. Interactions of trading strategies. Clustering of red blood cells in capillaries. Patterns of worm appendages. Shapes of galaxies. Effects of fires on ecosystems. Structure of stromatolites. Patterns of leaf stomata operation. Spatial spread of influenza in hospitals. Pedestrian traffic flow. Skin cancer development. Size distributions of companies. Microscopic origins of friction. And many, many more.

One of the key lessons of NKS is that even when a phenomenon appears complex, there may still be a simple underlying model for it. And to me one of the most interesting features of the applied NKS literature is that over the course of the decade typical successful models have been getting simpler and simpler—presumably as people get more confident in using the methods and ideas of NKS.

Of NKS-based models now in use, the vast majority are still based on cellular automata—the very first simple programs I studied back in the early 1980s. There's been some use of substitution systems, mobile automata and various graph-based systems. But cellular automata are still definitely the leaders. And beginning with my work on them in the early 1980s, there's been steady growth in their use for nearly 30 years. (Notably, the numbers of papers about them firmly overtook the number of papers about Turing machines around 1995.)

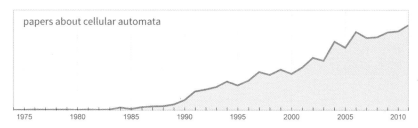

It's notable that the breakdown of cellular automaton papers is distinctly different from the breakdown of all NKS papers.

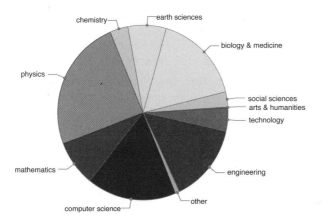

And it's a great testament to the importance of simple rules that even among the 256 simplest possible cellular automata, there've now been papers written about almost every single one of them. (The top rules, which also happen to be top in the NKS book, are rule 110, rule 30 and rule 90—which respectively show provable computation universality, high-quality randomness generation and additive nested structure capable of mathematical analysis).

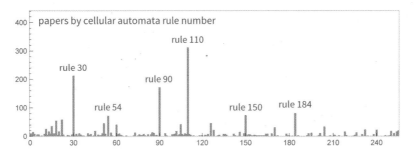

Many of these papers are applied NKS, using cellular automata as models (rule 184 for traffic flow, rule 90 for catalysis, rule 54 for phase transitions, etc.). But many are also pure NKS, studying the cellular automata for their own sake.

In a sense, each possible cellular automaton rule is a world unto itself. And over the course of the past decade, a whole variety of pockets of literature have developed around specific rules—typically using various computational and mathematical methods to identify and prove features of their behavior (nesting in the boundary of rule 30, blocks in rule 146, logic structures in rule 54, etc.). And in general, it continues to amaze me just how many things are still being discovered even about the very simplest possible cellular automaton rules—that I first studied 30 years ago.

One of my favorite activities when I worked on the NKS book was just to go out into the computational universe, and explore what's there. In a sense it's like quintessential natural science—but now concerned with exploring not stars and galaxies or biological flora and fauna, but instead abstract programs. Over the years I've developed a whole methodology for exploring the computational universe, and for doing computer experiments. And it's interesting to see the gradual spread of this methodology—rather obviously visible, for example, in all sorts of papers with pictures that look as if they could have come straight from the NKS book. (Most often, there are arrays of black-and-white rasters and the like; the more elaborate style of algorithmic diagrams in the NKS book are still much rarer—even though recent versions of Mathematica have made them easier to make.)

There's been an issue, though, in that raw explorations of the computational universe don't fit terribly well into current patterns of academia and academic publishing. We've tried to do what we can with our *Complex Systems* journal (published since 1986), but ultimately there are new venues and models needed, based more on structured knowledge and less on academic narrative. Soon after the NKS book came out, we did some experiments on developing an "Atlas of Simple Programs"—but we realized that to make this truly useful, we'd need something not so much like an atlas, but more like a computational knowledge engine. And of course, now with Wolfram|Alpha we have just that—and we're gradually adding lots of NKS knowledge to it:

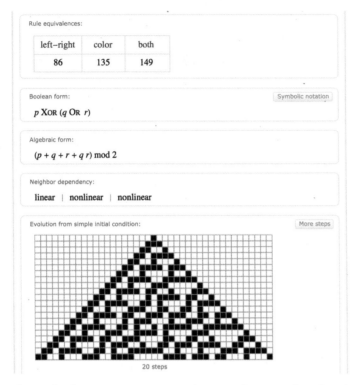

If one looks at pure NKS work over the past decade, the vast majority has been concerned with types of systems that I already discussed in some form or another in the book. Cellular automata remain the most common, but there are also recursive sequences, substitution systems, network systems, tag systems, and all sorts of others. And there've been some significant generalizations, and some new kinds of systems. Cellular automata with non-local rules, with memory, or on networks. Turing machines with skips. Iterated finite automata. Distance transform automata. Planar trinets. Generalized reversal-addition systems. And others.

Much of the systematic work that's been done on pure NKS has been by "professional scientists" who publish in traditional academic journals. But some of the most interesting and innovative work has actually been done by amateurs (most often people involved in some facet of the computer industry) who have the advantage of not having to follow the constraints of academic publishing—but at least for now have the disadvantage that there is no centralized venue for their contributions.

In pure NKS, one often starts from raw observations of the computational universe, but then moves on to detailed analysis, and to the formulation of increasingly general hard (at least for me) to tell if it makes sense (relations to Eastern religious traditions, unusual sensory experiences, etc.).

It's been fun over the past decade to watch NKS gradually make its way into popular culture. Whether it's in cartoon strips that pithily use some NKS idea. Or in fiction that uses NKS to theme some character. Or whether it's a cameo appearance of the physical NKS book—or an NKS piece of dialog—on a TV show.

There've been quite a few science fiction novels where NKS has been central. Sometimes there's a hero or villain who pursues NKS research or ideas. Sometimes NKS enters in the understanding of what a future governed by computation would be like. And sometimes there's some object whose operation or purpose can only be understood in NKS terms.

In a rather different direction, another major use of NKS has been in art. From my very earliest investigations of cellular automata, we've known that they can create rich and aesthetically pleasing visual images. But the publication of the NKS book dramatically accelerated the use of systems like cellular automata for artistic purposes. And in the past decade I've seen cellular automaton patterns artistically rendered in paint, knitting, mosaic, sticks, cake decoration, punched holes, wood blocks, smoke, water valves and no doubt all sorts of other media that I can't immediately even imagine.

In the interactive domain, there are countless websites and apps that run cellular automata and other NKS systems. Sometimes what's done is a fairly straightforward rendering of the underlying system—not unlike what's in the NKS book. And sometimes there is extensive artistic interpretation done. Sometimes what's built is used essentially for hobbyist NKS explorations, sometimes for art, and sometimes for pure entertainment.

In games—and movies—NKS systems have been extensively used both to produce detailed effects such as textures, and to handle for example collective behavior of large numbers of similar entities. NKS systems are also used to define overall rules for games or for characters in games.

NKS is not limited to visual form. And indeed NKS systems have been used extensively to generate audio—with probably the most ambitious experiment in this direction being our 2005 WolframTones project.

Of all responses to the NKS book, one of the consistently strongest has been from the architecture community. Many times I have heard how much architects value both the physical book and its illustrations, as well as the ideas about creation of form that it contains.

I don't know if there's any large actual building yet constructed from the simple rules of an NKS system. But I know quite a few have been planned. And there's also been significant work done on landscape architecture and urban planning—not to mention interior design—with NKS systems.

There've been many important applications of NKS in the sciences, humanities and arts over the past decade. But if there's one area where the effect of NKS has been emerging as most significant, it's technology. Just as the computational universe gives us an inexhaustible supply of new models, so also it gives us an inexhaustible supply of new mechanisms that we can harness for technology. And indeed in our own technology development over the past decade, we have made increasing use of NKS.

A typical pattern is that we identify some task, and then we search the computational universe for a simple program to carry out the task. Often this kind of mining of the computational universe seems like magic—and the programs we discover seem much cleverer than anything we would have ever come up with by our usual engineering methods of step-by-step construction.

Years ago we started using rule 30 as the pseudorandom generator in Mathematica—and more recently we've done a large-scale search to find other cellular automata that are slightly more efficient for this task. Over the years we've also mined the computational universe to find efficient hash codes, function evaluation algorithms, image filters, linguistics algorithms, visual layout methods and much, much more. Sometimes there's something known from pure NKS that immediately suggests some particular rule or type of system to use; more often we need to do a systematic search from scratch.

It's a little hard to track how NKS has been used in technology in the world at large. There are a handful of patents that refer to NKS, and a number of organizations have systematically sought my advice about using NKS—but most of what I've heard about technology uses of NKS has come from anecdotes and chance encounters.

Sometimes it's at least somewhat clear on the surface, because there's some visible NKS-like pattern to be seen—say in window shade arrangements for buildings, antenna designs, contact lens patterns, heat conduction networks, and so on.

But more often it's a case of "NKS inside", where there's some internal algorithm that's been found by exploring the computational universe. By now I know of a large number of examples. Modular robots. Mesh networks. Cryptography. Image processing. Network routing. Computer security. Concurrency protocols. Computer graphics. Algorithmic texture generation. Probabilistic processor designs. Games. Traffic light control. Voting schemes. Financial trading systems. Precursors to algorithmic drugs. And many more.

It's not like the academic literature, where (at least in theory) there are citations to be followed. But my strong impression is that the use of NKS in practical technology is starting to dwarf its use in basic research. In part this is just a sign of healthy development in the direction of applications.

But in part it is, I think, also a reflection of limitations in the current structure of academia and basic research. For university departments and research funding programs are typically organized according to disciplines that are many decades old—and are ill equipped to adapt to new directions like NKS that do not fit into their existing structures.

Still, over the past decade, there have sprung up around the world a good number of groups that are successfully carrying out basic NKS research. The story in each case is different. Particular leadership; special reasons for institutional or government support; particular ways to survive "interstitially" in an institution; and so on. But gradually more and more of these pockets are developing, and prospering.

When the NKS book appeared, I visited many universities and research labs—and talked to many university presidents, lab directors, and the like. A common response was excitement not only about the subject matter of NKS, but also about the social phenomenon that at talks and lunches about NKS, people who might have been in different parts of an institution for decades were actually meeting—and interacting—in ways they'd never done before.

For a while I thought it might make sense to use this enthusiasm to help launch strong NKS initiatives at particular existing institutions. But after some investigation, it became clear that this was not going to be any kind of quick process. It wasn't like at our company, where a significant new initiative can be launched in days. This was something that was going to take years—if not decades—to achieve.

And actually, I pretty much knew what to expect. Because I'd seen something very similar before. Back in the early 1980s, after I made my first discoveries about cellular automata, I realized that studying complex systems with simple rules was going to be an important area. And so I set about developing and promoting what I called "complex systems research" (I avoided the eventually-more-common term "complexity theory" out of a deference to the existing area of theoretical computer science by that name).

I soon started the first research center in the field (the Center for Complex Systems Research at the University of Illinois), and the first journal (*Complex Systems*). I encouraged the then-embryonic Santa Fe Institute to get into the field (which they did), and did my best to expand support for the field. But things went far too slowly for me—and in 1986 I decided that a better personal strategy would be to build the best possible tools I could (Mathematica), get the best possible personal environment (Wolfram Research), and then have the fun of doing the research myself. And indeed this is how I came to embark on *A New Kind of Science*, and ultimately to take things in a rather different direction.

So what happened to complexity? Here's a plot of the number of "complexity institutes" in the world by founding date (at least the ones we could find).

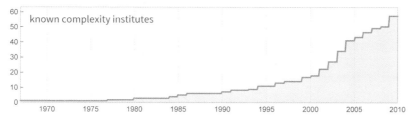

And here's their current geographic distribution (some excellent ones are doubtless missing!):

(Fully half of all complexity institutes we identified didn't tell us their founding dates. The institutes in operation before 1986 were originally rather different from what I called complex systems research, but have evolved to be much closer.)

So what does this tell us? The main thing as far as I'm concerned is that the development of institutions is slow and inexorable. Until we collected the data I had no idea there were now so many complexity institutes in the world. But it took close to three decades to get to this point. I think NKS is off to a quicker start. But it's not clear what one does to accelerate it, and it's still inevitably going to take a long time.

I might say that I think the intellectual development of NKS is going much better than it ever did for complexity. Because for NKS there's a core of pure NKS—as well as the NKS way of thinking—that's successfully being studied, and building up a larger and larger body of definite formal knowledge. In complexity, however, there was an almost immediate fragmentation into lots of different—and often somewhat questionable—application areas. Insofar as these applications followed traditional disciplinary lines, it was often easier to get institutional support for them. But with almost no emphasis on any kind of core intellectual structure, it's been difficult for coherent progress to be made.

What's exciting to see with NKS is that in addition to strong applications—especially in technology—there is serious investment in general abstract core development. Which means that independent of success in any particular application area, there's a whole intellectual structure that's growing. And such structures—like for example pure mathematics—have tended to have implications of great breadth, historically spanning millennia.

So how should people find out about NKS? Well, I put all the effort I did into writing *A New Kind of Science* to make that as easy as possible for as broad a range of people as possible. And I think that for the most part that worked out very well. Indeed, it's been remarkable over the past decade how many people I've run into—often in unexpected places—who seem to have read and absorbed the contents of the NKS book.

And to me what's been most interesting is how many different walks of life those people come from. Occasionally they're academics, with specific interests. More often they're other kinds of intellectually oriented people. Sometimes they're involved with computers; sometimes they're not. Sometimes they're highly educated; sometimes they're not. Sometimes they're young; sometimes they're old. And occasionally they turn out to be famous—often for things that at first seem to have no connection to interest in NKS.

I suppose for me one of the most satisfying things about the spread of NKS is seeing people get so much pleasure out of taking on the serious pursuit of NKS. I have a theory that for almost everyone there is a certain direction that's really their ideal niche. Too often people go through their lives without finding it. Or they live at a time in history when it simply does not exist. But when something like NKS comes along, there's a certain set of people for whom this is their thing: this is the direction that really fits them. I'm of course an example, and I feel very fortunate to have found NKS. But what's been wonderful over the past decade is to see all sorts of other people—at different stages in their lives—having that experience too.

There's much to do in spreading knowledge about NKS. I started the job with the NKS book. But by now NKS has found its way into plenty of textbooks and popular books. Our Demonstrations Project has lots of interactive demonstrations of NKS concepts. And there are other apps and websites too.

For me a major effort in NKS education has been our Summer School, held every year since 2003. We've had a diverse collection of outstanding students, and it's been invigorating each year to see the variety of NKS projects that they've been able to pursue. But I suppose if there's been one discovery for me from the Summer School, it's been what an important general educational foundation NKS provides.

One can study NKS for its own sake, just as one can study math or physics for their own sake. But like math or physics, one can also study NKS as a way to develop general patterns of thinking to apply all over the place. This works at the level of education for business executives—and it also works at much lower educational levels.

When I first saw people suggesting that young kids work out cellular automaton evolutions by hand on graph paper, I thought it was a little nutty. But then I realized that it's a great exercise in precision work, with rather satisfying results, that also teaches a kind of "pre-computer-science" idea of what algorithms are. Oh, and it can also be quite aesthetic, and even has direct connections with nature (yes, teachers ask us where to get patterned mollusc shells).

There have now been many courses at graduate, undergraduate and high-school levels that use the NKS book. And while I don't know everything that's happened, I know that there have at least been experiments with NKS at middle-school and elementary-school levels.

For me, one of the interesting things is that it's so easy for anyone to discover something original in NKS. At the beginning of our Summer School we always ask each student to find an "interesting" cellular automaton—and they inevitably come back with a rule that has never been seen before. The computational universe is so vast—and still so unexplored—that anyone can find their own part of it. And then it's a question of having a good methodology and being systematic to be able to discover something of real value.

When I look at the NKS book now, I'm pleased how well it has withstood the past 10 years. New things have been discovered, but they do not supersede what's in the book. And even with all the scrutiny and detailed study the book has received, not a single significant error has surfaced. And the pictures are direct and abstract enough to be in a sense timeless (much like a Platonic solid from ancient Egypt looks the same as one today). Of course, while the paper version of the book is elegant and classic—and seems to me a little ceremonial—the book is now much more read on the web and the iPad.

But what of the detailed content of the book? There are certainly plenty of academic papers that cite specific pages of the book. But as I look through the book, I cannot help but be struck by how much more there is to "mine" from it. In writing the book—with Mathematica at my side—it felt as if I had developed a kind of industrial-scale way of making discoveries. Thousands of them. That I discussed in the main part of the book, or put in tiny print and little diagrams into the notes at the back.

It's been interesting to see what's been absorbed, at what rate. Different fields seem to have different characteristic times. Art, for example, is very fast. Within months after the NKS book came out, there was significant activity among artists based on NKS concepts. No doubt because in art there's always excitement about the new.

The same was true for example in finance and trading. Where again there is a great premium for the new.

But in other fields, things were slower. The younger and more entrepreneurial the field, the faster things seemed to go. And the less set the conceptual framework for the field, the deeper the early absorption would be. In fields like mathematics and physics where there's a tradition of precise results, some basic technical understanding came quickly. But in these old and mature fields, the rate of real absorption seemed to be incredibly slow. And indeed, if I look at what I consider major gaps in the follow-up to the NKS book, they are centered around mathematics and physics—with a notable area being the work on the fundamental theory of physics in the book.

In a sense, ten years is a short time for the kind of development that I believe NKS represents. Indeed, in my own life so far, I have found that ten years is about the minimum time it takes me to really begin to come to terms with new frameworks for thinking about things. And certainly in the general history of science and of ideas, ten years is no time at all. Indeed, ten years after the publication of Newton's *Principia*, for example, almost none of the people who would make the next major advances had even been born yet.

And ten years after the publication of *A New Kind of Science*, I am personally quite satisfied with what I can see of the development of the field. Certainly there is far to go, but a solid start has been made, many important steps have been taken, and the inexorable progress of NKS seems assured.

# Living a Paradigm Shift: Looking Back on Reactions to *A New Kind of Science*

Published May 11, 2012

"You're destroying the heritage of mathematics back to ancient Greek times!" With great emotion, so said a distinguished mathematical physicist to me just after *A New Kind of Science* was published ten years ago. I explained that I didn't write the book to destroy anything, and that actually I'd spent all those years working hard to add what I hoped was an important new chapter to human knowledge. And, by the way—as one might guess from the existence of Mathematica—I personally happen to be quite a fan of the tradition of mathematics.

He went on, though, explaining that surely the main points of the book must be wrong. And if they weren't wrong, they must have been done before. The conversation went back and forth. I had known this person for years, and the depth of his emotion surprised me. After all, I was the one who had just spent a decade on the book. Why was he the one who was so worked up about it?

And then I realized: this is what a paradigm shift sounds like—up close and personal. I had been a devoted student of the history of science for many years, so I thought I knew the patterns. But it was different having it all unfold right around me.

I had been building up the science in the book for the better part of 20 years. And I had been amazed—almost shocked—at many of the things I'd discovered. And I knew that communicating it all to the world wouldn't be easy.

In the early years, I'd just done what scientists typically do, publishing papers in academic journals and giving talks at academic conferences. And that had gone very well. But after I built Mathematica, I started being able to discover things faster and faster. I had a great time. And pretty soon I had material for many tens—if not hundreds—of academic papers. And what's more, the things I was discovering were starting to fit together, and give me a whole new way of thinking.

What was I going to do with all this? I suppose I could have just kept it all to myself. After all, by that time I was the CEO of a successful company, and certainly didn't make my living by publishing research. But I thought what I was doing was important, and I really liked the idea of giving other people the opportunity to share the enjoyment of the things I was discovering. So I had to come up with some way to communicate it all. Publishing lots of piecemeal academic papers in the journals of dozens of fields wasn't going to work. And instead it seemed like the best path was just to figure out as much as I could, and then present it to the world all together in a coherent way. Which in practice meant I had to write a book.

Back in 1991 I thought it might take me a year, maybe two, to do this. But I kept on discovering more and more. And in the end it took nearly 11 years before I finally finished what I had planned for *A New Kind of Science*.

But whom was the book supposed to be for? Given all the effort I had put in, I figured that I should make it as widely accessible as possible. I'm sure I could have invented some elaborate technical formalism to describe things. But instead I set myself the goal of explaining what I'd discovered using just plain language and pictures. And as it turned out, countless times doing this helped me clarify my own thinking. But it also made it conceivable for immensely more people to be able to read and understand the book.

I knew full well that all this was very different from the usual pattern in science. Most of the time, front-line research gets described first in academic papers written for experts. And by the time it gets into books—and especially broadly accessible ones—it's not new, and it's usually been very watered down. But in my case, many years of front-line discoveries would first get described in a broadly accessible book.

Even in the preface I wrote for the book, I expressed concern about how specialist scientists would react to this. But my personal decision was that it was worth it. And when the book came out, it indeed for the most part worked out spectacularly. Most important as far as I was concerned was that a huge spectrum of people were able to read and understand the book. And in fact lots of people specifically thanked me for writing the book so it was accessible to them.

Many specialist scientists were also highly enthusiastic about the book. But much as I had expected, there was a certain component who just assumed that anything presented in a bestselling book couldn't really be important new science—and pretty much stopped there.

And then there were others for whom the book just seemed irrelevant; they were happy with their ways of doing and thinking about science, and they weren't interested—at least at that time—in any particular injection of new ideas.

But what about people whose work was in some way or another directly connected to issues discussed in the book? I have to say that by and large I expected a very positive reaction from such people. After all, I had put all this meticulous work into studying things that they were interested in—and I believed I had come up with some exciting results. And what's more, I personally knew many of these people—and lots of them had benefited greatly from my efforts to develop the field of complex systems research some 15 or so years earlier.

So discussions like the one I described at the beginning of this post at first came as something of a shock. Of course, I had spent quite a few years as an academic, so I was well aware of the petty bickering and backstabbing endemic to that profession. But this was something different: this was people who were somehow deeply upset by what I was doing.

Some of the more sophisticated and forthright of them were pretty explicit with me, at least in person. Typically there was a surface reason for their reaction, and a deeper reason. Sometimes the surface reason related to content, sometimes to form. Those who discussed content fell into two main groups. The first—particularly populated by physicists—said that their immediate reason for being upset was basically that "If what you're doing is right, we've spent our whole careers barking up the wrong tree". The second group—particularly populated by people who'd studied areas related to complexity—basically said: "If people buy into what you've done, it'll overshadow everything we've done".

To me what was perhaps most striking was that these reactions were often coming from some of the best-established and most senior people in their fields. I suppose at some level it was flattering, but I had certainly not expected this kind of insecurity—not least because I thought it was completely unfounded. Yes, I believed new approaches were needed—that's why I'd spent so many years developing them. But I saw what I had done as complementing and adding to what had been done before, not replacing or overturning it.

And then there was the matter of form. "You've done something that's academic-like, but you haven't played by academic rules." It was true: I wasn't an academic and I wasn't operating according to the constraints of academia; I was just trying to invent the best possible ways to do things, given my resources and the discoveries I was making.

A typical issue that came up was how the book was vetted or checked. In academia, there's the idea that "peer review" is the ultimate method of checking anything. And perhaps in a world where everyone has infinite time, and nobody operates according to their own self-interest, this might be true. But in reality, peer review is fraught with error, often quite corrupt, and even in the best case strongly biased toward avoiding new ideas and maintaining the status quo. And for a piece of work as large, broad and complex as *A New Kind of Science*, even the basic mechanics of it seemed completely impractical.

So what did I do instead? It was a big exercise in perfectionism. First, we built a large system for automated testing, modeled on what we'd developed over the course of many years for Mathematica. And then we developed a process for getting experts in different areas to look at every page of the book, checking as far as possible every detail. So how did it work out? Impressively well. For even now, 10 years later, after every page of the book has been read and scrutinized by huge numbers of people, and every computational result has been reproduced many times, in all the 1280 pages of the book no errors much beyond simple typos have come to light.

One of the many challenges with the book was the practicality of actually printing and publishing it. Early on, I had hoped that one of the large publishers who wanted to publish the book would be able to handle it. But after awhile it became clear that their production methods and business models could not realistically handle the level of visual quality that the content of the book required. And so, somewhat reluctantly at first, I decided to have Wolfram Media do the publishing instead.

This certainly allowed the book to be printed at higher quality, and to be sold more cheaply. But the setup definitely seemed not to please some academics—particularly, I suspect, because it made clear that the book was simply beyond the reach of any academic network, however powerful that network might be in the academic world. Even if a couple of times I did hear things like "I'm so shocked about [some aspect of your book] that I'm going to campaign my university not to use Mathematica".

If one looks at a standard academic paper, one of its prominent elements is always a list of references—in principle a list of authorities for statements in the paper. But *A New Kind of Science* has no such list of references. And to some academics this seemed absolutely shocking. What was I thinking? I always consider history important—both for giving credit and for letting one better understand the context of ideas. And when I wrote *A New Kind of Science*, I resolved that rather than just throwing in disembodied references, I would actually do the work of trying to unravel and explain the detailed histories of things.

And the result was that of the nearly 300,000 words of notes at the back of the book, a significant fraction are about history. I did countless hours of (often fascinating) primary interviews and went through endless archives—and in the end was rather proud of the level of historical scholarship I managed to achieve. And when it came to traditional references I figured that rather than using yet more printed pages, I should just include in the notes appropriate names and keywords, from which anyone—even with the state of web search in 2002—could readily find whatever primary literature they wanted, at greater depth and more conveniently than from lists of journal page numbers.

When *A New Kind of Science* came out, I kept on hearing complaints that it didn't refer to this or that person or piece of work. And each time I would check. And to my frustration, in almost every case it was right there in the book—with a whole extended historical story. And it wasn't as if when people actually read it, they disagreed with the history I'd written. Indeed, to the contrary, many times people told me they were impressed at how accurate and balanced my account was—and often that they'd learned new things even about pieces of history in which they were personally involved.

So why were people complaining? I think it was somehow just disorienting for academics not to be able to glance down a definite "references section"—and see papers they'd authored or otherwise knew. But I'm pretty sure it was more an emotional than a functional issue. And as one indication, after the book came out we did the experiment of putting on the web—in standard academic reference format—the list of the 2600 or so books that I'd used in writing *A New Kind of Science*. And from our web statistics we know that vastly fewer people used this than for example the online version of even one chapter's worth of historical notes. (Even so, as a matter of completeness, I'm hoping one day to link all my archives of papers to the online book.)

I suppose another feature of the book that did not endear it to some academics was the very intensity of positive reaction that accompanied its release. Within days there were hundreds of articles in the media describing the ideas in the book—with journalists often doing an impressive job of understanding what the book had to say. And then, mostly slightly later, reviews started to appear. Some were detailed and well reasoned;

others were quite rushed, and often seemed mainly to be emotional responses—probably more based on reading earlier reviews than on the reading of the actual 1280-page book itself. And after such a positive initial wave of media attention, later (often "catch up") coverage inevitably tended to swing to the more negative.

When the book came out, I had all the reviews we could find diligently archived. And I always intended at some point to systematically read them. But somehow a decade has gone by, and I have not done so. And as I write this post, I have on my desk a daunting pile of printed copies of reviews, as thick as the book itself. But back when they were archived, for reasons I don't now know, each review was at least put into a 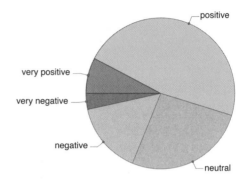 "star-rating" category, which we can now use to make a pie chart. And while I'm not sure just how much these statistics really mean, it is perhaps interesting that positive—or at least neutral—reviews overall significantly outweighed negative ones.

So what about the negative reviews? There are certainly some colorful quotes in them. "Why has this undoubtedly brilliant, worthily successful man written such a silly book?... I think it... likely that the book will be forgotten in a few months." "There's a tradition of scientists approaching senility to come up with grand, improbable theories. Wolfram is unusual in that he's doing this in his 40s." "Is this stuff really that important? Well... maybe. Frankly, I doubt it." "After looking at hundreds of Wolfram's pictures, I felt like the coal miner in one of the comic sketches in *Beyond the Fringe*, who finds the conversation down in the mines unsatisfying: 'It's always just "Hallo, 'ere's a lump of coal."'" "With extreme hubris, Wolfram has titled his new book on cellular automata 'A New Kind of Science'. But it's not new. And it's not science." "It was not the first time the names Wolfram and Newton have been mentioned in the same breath, and I suppose it might be taken as further evidence of an ego bursting all bounds." And, perhaps my favorite, a whole review simply titled "A Rare Blend of Monster Raving Egomania and Utter Batshit Insanity".

Realistically, much of the space in these reviews was not devoted to discussing the actual content of the book. High on the list of issues they discussed was that the book had not gone through an academic peer review process, and so could "not be considered an academic work" (not that it was meant to be). Then there were the complaints about the absence of explicit lists of references—often rather misleadingly with no mention of all the detailed historical notes, or perhaps a grudging comment that they were in too small a font.

Another common complaint was that the book was somehow just too grandiose. And for sure, any book with a title like "A New Kind of Science" runs the risk of being characterized that way. To be clear, I believed—and very much still believe—that what's in *A New Kind of Science* is very important. In presenting it, though, I suppose I could have somehow tried to hide this. But I was fairly sure that doing so would have a bad effect on people's ability to understand what was in the book.

The issue is quite familiar to those of us who have written lots of documentation for computer systems: if you have big ideas to communicate, you have to prime people for them—or they inevitably get confused. Because if people think something is a small idea, they'll try to understand it by straightforwardly extending what they already know. And when that doesn't work, they'll just be confused. On the other hand, if you communicate up front that something is big and important, then people will make the effort to understand it on its own terms—and will much more readily be able to place and absorb it. And so—well aware of the potential for being accused of grandiosity—I made the decision that it was better for the science if I was explicit about what I thought was important, and how important I thought it was.

Looking through reviews, there are some other common themes. One is that *A New Kind of Science* is a book about cellular automata—or worse, about the idea (not in fact suggested in the book at all) that our whole universe is a giant cellular automaton. For sure, cellular automata are great, visually strong examples for lots of phenomena I discuss. But after about page 50 (out of 1280), cellular automata no longer take center stage—and notably are not the type of system I discuss in the book as possible models for fundamental physics.

Another theme in some reviews is that the ideas in the book "do not lead to testable predictions". Of course, just as with an area like pure mathematics, the abstract study of the computational universe that forms the core of the book is not something which in and of itself would be expected to have testable predictions. Rather, it is when the methods derived from this are applied to systems in nature and elsewhere that predictions can be made. And indeed there are quite a few of these in the book (for example about repeatability of apparent randomness)—and many more have emerged and successfully been tested in work that's been done since the book appeared.

Interestingly enough, the book actually also makes abstract predictions—particularly based on the Principle of Computational Equivalence. And one very important such prediction—that a particular simple Turing machine would be computation universal—was verified in 2007.

There are reviews by people in specific fields—notably mathematics and physics—that in effect complain that the book does not follow the methodology of their field, which is of course why the book is titled "A New Kind of Science". There are reviews by various academics with varied "it's been done before" claims. And there are a few reviews with specific technical complaints, about definitions or about phenomena like the emergence of quantum effects from essentially deterministic systems. Sometimes the issues brought up are interesting. But so far as I know not a single review brought up any specific relevant factual issue that wasn't in some way already addressed in the book.

And reading through negative reviews, the single most striking thing to me is how shrill and emotional most of them are. Clearly there's more going on than just what's being said. And this is where the paradigm shift phenomenon comes in. Most people get used to doing science (or other things) in some particular way. And there's a natural tendency to want to just go on doing things the same way. And there's no issue with that for people whose subject matter is sufficiently far away—and who can successfully say, "I just don't care about your new kind of science". But for people whose subject matter is closer, that doesn't work. And that's when the knives really come out.

I have to say that to me (as I discussed in the previous chapter) the progress of NKS seems quite inexorable, and unavoidable—and indeed a decade after the publication of the book, seems to be progressing well. But I think some of the reviewers of *A New Kind of Science* convinced themselves that if what they wrote was negative enough they could derail things, and maybe allow their old directions and paradigms to continue unperturbed. And perhaps the mathematical physicist mentioned at the beginning of this post expressed their attitude most clearly when he said in our conversation: "You're one of the most brilliant people I know… but you should keep out of science".

There's lots of analysis that could be done of the dynamics of opinions about *A New Kind of Science*. In 2002, there were fewer venues than today for public comments to be made. But I suspect that enough existed that it would be possible to piece together much of what happened. And I think it makes a fascinating study in the history of science.

I suppose I am myself by nature a positive person. And no doubt that is a necessary trait if one is going to do the kinds of large projects to which I have devoted most of my life. I am also at this point someone who is much more interested in just doing things than in other people's assessments of what I do. No doubt others would find attacks on as important a personal project as *A New Kind of Science* dispiriting. But I have to say that first and foremost my reaction was one of scientific interest. Having studied so much history about paradigm shifts, I found it fascinating to be right in the middle of one myself.

I certainly wondered what one could predict from the dynamics of what was going on. And here history has some interesting lessons. For it suggests that perhaps the single best predictor of good long-term outcomes from potential paradigm shifts is how emotional people get about them at the beginning. So for NKS, all that early turbulence in the end just helps fuel my optimism for its long-term importance and success. And a decade out, not least with everything my previous chapter discussed, things indeed seem to be well on their way.

# At 10 Years: Looking to the Future of *A New Kind of Science*

*Published May 14, 2012*

Today ten years have passed since *A New Kind of Science* ("the NKS book") was published. But in many ways the development that started with the book is still only just beginning. And over the next several decades I think its effects will inexorably become ever more obvious and important.

Indeed, even at an everyday level I expect that in time there will be all sorts of visible reminders of NKS all around us. Today we are continually exposed to technology and engineering that is directly descended from the development of the mathematical approach to science that began in earnest three centuries ago. Sometime hence I believe a large portion of our technology will instead come from NKS ideas. It will not be created incrementally from components whose behavior we can analyze with traditional mathematics and related methods. Rather it will in effect be "mined" by searching the abstract computational universe of possible simple programs.

And even at a visual level this will have obvious consequences. For today's technological systems tend to be full of simple geometrical shapes (like beams and boxes) and simple patterns of behavior that we can readily understand and analyze. But when our technology comes from NKS and from mining the computational universe there will not be such obvious simplicity. Instead, even though the underlying rules will often be quite simple, the overall behavior that we see will often be in a sense irreducibly complex.

So as one small indication of what is to come—and as part of celebrating the first decade of *A New Kind of Science*—starting today, when Wolfram|Alpha is computing, it will no longer display a simple rotating geometric shape, but will instead run a simple program (currently, a 2D cellular automaton) from the computational universe found by searching for a system with the right kind of visually engaging behavior.

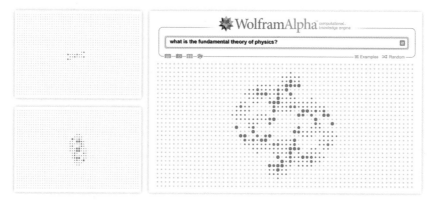

This doesn't look like the typical output of an engineering design process. There's something much more "organic" and "natural" about it. And in a sense this is a direct example of what launched my work on *A New Kind of Science* three decades ago. The traditional mathematical approach to science has had great success in letting us understand systems in nature and elsewhere whose behavior shows a certain regularity and simplicity. But I was interested in finding ways to model the many kinds of systems that we see throughout the natural world whose behavior is much more complex.

And my key realization was that the computational universe of simple programs (such as cellular automata) provides an immensely rich source for such modeling. Traditional intuition would have led us to think that simple programs would always somehow have simple behavior. But my first crucial discovery was that this is not the case, and that in fact even remarkably simple programs can produce extremely complex behavior—that reproduces all sorts of phenomena we see in nature.

And it was from this beginning—over the course of nearly 20 years—that I developed the ideas and results in *A New Kind of Science*. The book focused on studying the abstract science of the computational universe—its phenomena and principles—and showing how this helps us make progress on a whole variety of problems in science. But from the foundations laid down in the book much else can be built—not least a new kind of technology.

This is already off to a good start, and over the next decade or two I expect dramatic progress in the application of NKS to all sorts of technology. In a typical case, one will start from some objective one wants to achieve. Then, either through knowledge of the basic science of the computational universe, or by some kind of explicit search, one will find a system that achieves this objective—often in ways no human would ever imagine or come up with. We have done this countless times over the years for algorithms used in Mathematica and Wolfram|Alpha. But the same approach applies not just to programs implemented in software, but also to all kinds of other structures and processes.

Today our technological world is full of periodic patterns and other simple forms. But rarely will these ultimately be the best ways to achieve the objectives for which they are intended. And with NKS, by mining the computational universe, we have access to a much broader set of possibilities—which to us will typically look much more complex and perhaps random.

How does this relate to the kinds of patterns and forms that we see in nature? One of the discoveries of NKS is that nature samples a broader swath of the computational universe than we reach with typical methods of mathematics or engineering. But it too is limited, whether because natural selection tends to favor incremental change, or because some physical process just follows one particular rule. But when we create technology, we are free to sample the whole computational universe—so in a sense we can greatly generalize the mechanisms that nature uses.

Some of the consequences of this will be readily visible in the actual forms of technological objects we use. But many more will involve internal structures and processes. And here we will often see the consequences of a central discovery of NKS: the Principle of Computational Equivalence—which implies that even when the underlying rules or components of a system are simple, the behavior of the system can correspond to a computation that is essentially as sophisticated as anything. And one thing this means is that a huge range of systems are capable in effect not just of acting in one particular way, but of being programmed to act in almost arbitrary ways.

Today most mechanical systems we have are built for quite specific purposes. But in the future I have no doubt that with NKS approaches, it will for instance become common to see arbitrarily "programmable" mechanical systems. One example I expect will be modular robots consisting of large numbers of fairly simple and probably identical elements, in which almost any mechanical action can be achieved by an appropriate sequence of small-scale motions, typically combined in ways that were found by mining the computational universe.

Similar things will happen at a molecular level too. For example, today we tend to have bulk materials that are either perfect periodic crystals, or have atoms arranged in a random amorphous way. NKS implies that there can also be "computational materials" that are grown by simple underlying rules, but which end up with much more elaborate patterns of atoms—with all sorts of bizarre and potentially extremely useful properties.

When it comes to computing, we might think that to have a system at a molecular scale act as a computer we would need to find microscopic analogs of all the usual elements that exist in today's electronic computers. But what NKS shows us is that in fact there can be much simpler elements—more readily achievable with molecules—that nevertheless support computation, and for which the effort of compiling from current traditional forms of computation is not even too great.

An important application of these kinds of ideas is in medicine. Biology is essentially the only existing example where something akin to molecular-scale computation already occurs. But existing drugs tend to operate only in very simple ways, for example just binding to a fixed molecular target. But with NKS methods one can expect instead to create "algorithmic drugs", that in effect do a computation to determine how they should act—and can also be programmable for different cases.

NKS will also no doubt be important in figuring out how to set up synthetic biological organisms. Many processes in existing organisms are probably best understood in terms of simple programs and NKS ideas. And when it comes to creating new biological mechanisms, NKS methods are the obvious way to take underlying molecular biology and find schemes for building sophisticated functionality on the basis of it.

Biology gives us ways to create particular kinds of molecular structures, like proteins. But I suspect that with NKS methods it will finally be possible to build an essentially universal constructor, that can in effect be programmed to make an almost arbitrary structure out of atoms. The form of this universal constructor will no doubt be found by searching the computational universe—and its operation will likely be nothing close to anything one would recognize from traditional engineering practice.

An important feature of NKS methods is that they dramatically change the economics of invention and creativity. In the past, to create or invent something new and original has always required explicit human effort. But now the computational universe in effect gives us an inexhaustible supply of new, original material. And one consequence of this is that it makes all sorts of mass customization broadly feasible.

There are many immediate examples of this in art. WolframTones did it for simple musical pieces. One can also do it for all sorts of visual patterns—perhaps ever changing, and selected from the computational universe and then grown to fit into particular spatial or other constraints. And then there is architecture. Where one can expect to discover in the computational universe new forms that can be used to create all sorts of structures. And indeed in the future I would not be surprised if at first the most visually obvious everyday examples of NKS were forms of things like buildings, their dynamics, decoration and structure.

Mass production and the legacy of the industrial revolution have led to a certain obvious orderliness to our world today—with many copies of identical products, precisely repeating processes, and so on. And while this is a convenient way to set things up if one must be guided by traditional mathematics and the like, NKS suggests that things could be much richer. Instead of just carrying out some processes in a precisely repeating way, one computes what to do in each case. And putting together many such pieces of computation the behavior of the system as a whole can be highly complex. And finding the correct rules for each element—to achieve some set of overall objectives—is no doubt best done by studying and searching the computational universe of possibilities.

Viewed from the outside, some of the best evidence for the presence of our civilization on Earth comes from the regularities that we have created (straight roads, things happening at definite times, radio carrier signals, satellite orbits, and so on). But in the future, with the help of NKS methods, more and more of these regularities will be optimized out. Vehicles will move in optimized patterns, radio signals will be transferred in complicated sequences of local hops... and even though the underlying rules may be simple, the actual behavior that is seen will look highly complex—and much more like all sorts of systems in physics and elsewhere that we already see in nature.

There are other—more abstract—situations where computation and NKS ideas will no doubt become increasingly important. One example is in commerce. Already there is an increasing trend toward algorithmic pricing. Increasingly commercial terms and contracts of all kinds will be stated in computational terms. And then—a little like a market of algorithmic traders—there will be what amounts to an NKS issue of what the overall consequences of many separate transactions will be. And again, finding the appropriate rules for these underlying transactions will involve understanding and searching the computational universe—and presumably various kinds of mass customization, that eventually make concepts like money as a simple numerical quantity quite obsolete.

Future schemes for such things as auctions and voting may also perhaps be mined from the computational universe, and as a result may be mass customized on demand. And, more speculatively, the same might be true for future corporate or political organizational structures. Or for example for mechanisms for social and other human networks.

In addition to using NKS in "technology mode" as a way to create things, one can also use NKS in "science mode" as a way to model and understand things. And typically the goal is to find in the computational universe some simple program whose behavior captures the essence of whatever system or phenomenon one is trying to analyze. This was an important focus of the NKS book, and has been a major theme in the past decade of NKS research. In general in science it has been difficult to come up with new models for things. But the computational universe is an unprecedentedly rich source—and I would expect that before long the rate of new models derived from it will come to far exceed all those from traditional mathematical and other sources.

An important trend in today's world is the availability of more and more data, often collected with automated sensors, or in some otherwise automated way. Often—as we see in many areas of Wolfram|Alpha or in experiments on personal analytics—there are tantalizing regularities in the data. But the challenge that now exists is to find good models for the data. Sometimes these models are important for basic science; more often they are important for practical purposes of prediction, anomaly detection, pattern matching and so on.

In the past, one might find a model from data by using statistics or machine learning in effect to fit parameters of some formula or algorithm. But NKS suggests that instead one should try to find in the computational universe some set of underlying rules that can be run to simulate the essence of whatever generates the data. At present, the methods we have for finding models in the computational universe are still fairly ad hoc. But in time it will no doubt be possible to streamline this process, and to develop some kind of highly systematic methodology—a rough analog of the historical progression from calculus to statistics.

There are many areas where it is clear that NKS models will be important—perhaps because the phenomenon being modeled is too complex for traditional approaches, or perhaps because, as is becoming so common in practice, the underlying system has elements that are specifically set up to be computational.

One area where NKS models seem likely to be particularly important is medicine. In the past, most disorders that medicine successfully addressed were fundamentally either structural or chemical. But today's most important challenge areas—like aging, cancer, immune response and brain functioning—all seem to be associated more with large-scale systems containing many interacting parts. And it certainly seems plausible that the best models for these systems will be based on simple programs that exist in the computational universe.

In recent times, medicine has slowly been becoming more quantitative. But somehow it is still always based on small collections of numbers, that lead to a small set of possible diagnoses. But between the coming wave of automated data acquisition, and the use of underlying NKS models, I suspect that the future of medicine will be more about dynamic computation than about specific discrete diagnoses. But even given a good predictive model of what is going on in a particular medical situation, it will still often be a challenge to figure out just what intervention to make—though the character of this problem will no doubt change when algorithmic drugs and computational materials exist.

What would be the most spectacular success for NKS models? Perhaps models that lead to an understanding of aging, or cancer. Perhaps more accurate models for social or economic processes. Or perhaps a final fundamental theory of physics.

In the NKS book, I started looking at what might be involved in finding the underlying rules for our physical universe out in the computational universe. I developed some network-based models that operate in a sense below space and time, and from which I was already able to derive some surprisingly interesting features of physics as we know it. Of course, we have no guarantee that our physical universe has rules that are simple enough to be found, say, by an explicit search in the computational universe. But over the past decade I have slowly been building up the rather large software and analysis capabilities necessary to mount a serious search. And if successful, this will certainly be an important piece of validation for the NKS approach—as well as being an important moment for science in general.

Beyond science and technology, another important consequence of a new worldview like NKS is the effect that it can have on everyday thinking. And certainly the mathematical approach to science has had a profound effect on how we think about all kinds of issues and processes. For today, whether we're talking about business or psychology or journalism, we end up using words and ideas—like "momentum" and "exponential"—that come directly from this approach. Already there are analogs from NKS that are increasingly used—like "computationally irreducible" and "intrinsically random". And as such concepts become more widespread they will inform thinking about more and more things—whether it's describing the operation of an organization, or working out what could conceivably be predictable for purposes of liability.

Beyond everyday thinking, the ideas and results of NKS will also no doubt have increasing influence on many areas of philosophical thinking. In the past, most of the understanding for what science could contribute to philosophy came from the mathematical approach to science. But now the new concepts and results in NKS in a sense provide a large number of new "raw facts" from which philosophy can operate.

The principles of NKS are important not only at an intellectual level, but also at a practical level. For they give us ideas about what might be possible, and what might not. For example, the Principle of Computational Equivalence in effect implies that there can be nothing general and abstract that is special about intelligence, and that in effect all its features must just be reflections of computation. And it is this that made me realize soon after the NKS book appeared that my long-term goal of making knowledge broadly computable might be achievable "just with computation"—which is what led me to embark on the Wolfram|Alpha project.

I have talked elsewhere about some of the consequences of the principles of NKS for the long-range future of the human condition. But suffice it to say here that we can expect an increasing delegation of human intellectual activities to computational systems—but with ultimate purposes still of necessity defined by humans and the history of human culture and civilization. And perhaps the place where NKS principles will enter most explicitly is in making future legal and other distinctions about what really constitutes responsibility, or a mind, or a memory as opposed to a computation.

As we look at the future of history, there are some inexorable trends, and then there are some wild cards. If we find the fundamental theory of physics, will we be able to hack it to achieve something like instantaneous travel? Will we find some key principle that lets us reverse aging? Will we be able to map memories directly from one brain to another, without the intermediate step of language? Will we find extraterrestrial intelligence? About all these questions, NKS has much to say.

If we look back at the mathematical approach to science, one of its societal consequences has been the injection of mathematics into education. To some extent, a knowledge of mathematical principles is necessary to interact with the world as it exists today. It is also an important foundation for understanding fields that have made serious use of the mathematical approach to science. And certainly learning mathematics to at least some level is a convenient way to teach precise structured thinking in general.

But I believe NKS also has much to contribute to education. At an elementary level, it can be viewed as a kind of "pre-computer science", introducing fundamental notions of computation in a direct and often visual way. At a more sophisticated level, NKS provides a conceptual framework for understanding the foundations of many computational fields. And even from what I have seen over the past decade, education about NKS—a little like physics before it—seems to provide a powerful springboard for people entering all sorts of modern areas.

What about NKS research? There is much to be done in the many applications of NKS. But there is also much to be done in pure NKS—studying the basic science of the computational universe. The NKS book—and the decade of research that has followed it—has only just begun to scratch the surface in exploring and investigating the vast range of possible simple programs. The situation is in some ways a little like in chemistry—where there are an infinite variety of possible chemical compounds each with their own features, that can be studied either for their own sake, or for the purpose of inferring general principles, or for diverse potential applications. And where even after a century or more, only a small part of what is possible has been done.

In the computational universe it is quite remarkable how much can be said about almost any simple program with nontrivial behavior. And the more one knows about a given program, the more potential there is to find interesting applications of it, whether for modeling, technology, art or whatever. Sometimes there are features of programs that can be almost arbitrarily difficult to determine. But sometimes they can be important. And so, for example, it will be important to get more evidence for (or against) the Principle of Computational Equivalence by trying to establish computation universality for a variety of simple programs (rule 30 would be a particularly important achievement).

As more is done in pure NKS, so its methodologies will become more streamlined. And for example there will be ever clearer principles and conventions for what constitutes a good computer experiment, and how the results of investigations on simple programs should be communicated. There are fields other than NKS—notably mathematics—where computer experiments also make sense. But my guess is that the kind of exploratory computer experimentation that is a hallmark of pure NKS will always end up largely classified as pure NKS, even if its subject matter is quite mathematical.

If one looks at the future of NKS research, an important issue is how it is structured in the world. Some part of it—like for mathematics—may be driven by education. Some part may be driven by applications, and their commercial success. But in the long term just how the pure basic science of NKS should be conducted is not yet clear. Should there be prizes? Institutions? Socially oriented value systems? As a young field NKS has the potential to take some novel approaches.

For an intellectual framework of the magnitude of NKS, a decade is a very short time. And as I write this post, I realize anew just how great the potential of NKS is. I am proud of the part I played in launching NKS, and I look forward to watching and participating in its progress for many years to come.

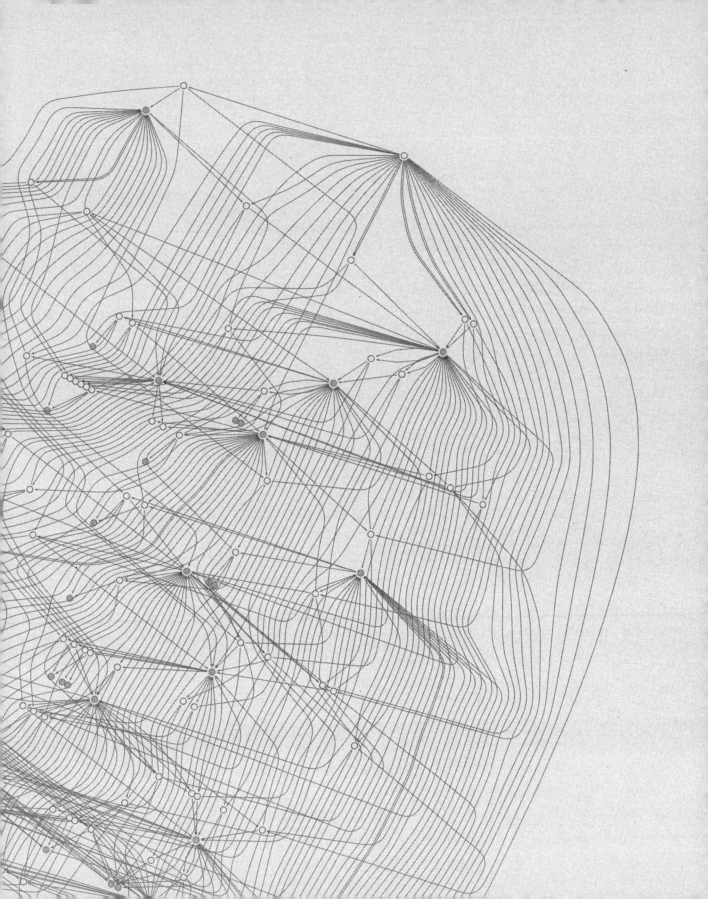

# The Making of
# *A New Kind of Science*

*Published May 13, 2022*

### I Think I Should Write a Quick Book...

In the end it's about five and a half pounds of paper, 1280 pages, 973 illustrations and 583,313 words. And its creation took more than a decade of my life. Almost every day of my thirties, and a little beyond, I tenaciously worked on it. Figuring out more and more science. Developing new kinds of computational diagrams. Crafting an exposition that I wrote and rewrote to make as clear as possible. And painstakingly laying out page after page of what on May 14, 2002, would be published as *A New Kind of Science*.

I've written before (even in the book itself) about the intellectual journey involved in the creation of *A New Kind of Science*. But here I want to share some of the more practical "behind the scenes" journey of the making of what I and others usually now call simply "the NKS book". Some of what I'll talk about happened twenty years ago, some more like thirty years ago. And it's been interesting to go back into my archives (and, yes, those backup tapes from 30 years ago were hard to read!) and relive some of what finally led to the delivery of the ideas and results of *A New Kind of Science* as truckloads of elegantly printed books with striking covers.

It was late 1989—soon after my 30th birthday—when I decided to embark on what would become *A New Kind of Science*. And at first my objective was quite modest: I just wanted to write a book to summarize the science I'd developed earlier in the 1980s. We'd released Version 1.0 of Mathematica (and what's now the Wolfram

Language) in June 1988, and to accompany that release I'd written what had rapidly become a very successful book. And while I'd basically built Mathematica to give me the opportunity to do more science, my thought in late 1989 was that before seriously embarking on that, I should spend perhaps a year and write a book about what I already knew, and perhaps tie up a few loose ends in the process.

My journey in science began in the early 1970s—and by the time I was 14 I'd already written three book-length "treatises" about physics (though these wouldn't see the light of day for several more decades). I worked purely on physics for a number of years, but in 1979 this led me into my first big adventure in technology—thereby starting my (very productive) long-term personal pattern of alternating between science and technology (roughly five times so far). In the early 1980s—back in a "science phase"—I was fortunate enough to make what remains my all-time favorite science discovery: that in cellular automaton programs even with extremely simple rules it's possible to generate immense complexity. And from this discovery I was led to a series of results that began to suggest what I started calling a general "science of complexity".

By the mid-1980s I was quite well positioned in the academic world, and my first thought was to try to build up the study of the "science of complexity" as an academic field. I started a journal and a research center, and collected my papers in a book entitled *Theory and Applications of Cellular Automata* (later reissued as *Cellular Automata and Complexity*). But things developed slowly, and eventually I decided to go to "plan B"—and just try to create the tools and environment that I would need to personally push forward the science as efficiently as possible.

The result was that in late 1986 I started the development of Mathematica (and what's now the Wolfram Language) and founded Wolfram Research. For several years I was completely consumed with the challenges of language design, software development and CEOing our rapidly growing company. But in August 1989 we had released Mathematica 1.2 (tying up the most obvious loose ends of Version 1.0)—and with the intensity of my other commitments at least temporarily reduced, I began to think about science again.

*The Mathematica Book* had been comparatively straightforward and fast for me to write—even as a "side project" to architecting and developing the system. And I imagined that it would be a somewhat similar experience writing a book explaining what I'd figured out about complexity.

My first working title was *Complexity: An Introduction to the Science of Complex Phenomena*. My first draft of a table of contents, from November 1989, begins with "A Gallery of Complex Systems" (or "The Phenomenon of Complexity"), and continues through nine other chapters, capturing some of what I then thought would be important (and in most cases had already studied):

```
contents      Fri Nov 17 00:32:20 1989        1

Complexity: An Introduction to the Science of Complex Phenomena

1. A Gallery of Complex Systems / The Phenomenon of Complexity
      Examples of complex systems and processes
            Physics  - snowflakes, fluids; ferrofluids;
                       dendritic growth
            Chemistry - Zhabotinsky
            Biology  - pigmentation patterns; morphogenesis in general;
                       structure of plants; neurons;
                       wood grain;
            Math - distribution of primes etc.; CA tapestry stuff;
                   3x+1 problem; Penrose tiling;
                   Mandelbrot set; Mod[x^2+y^3, 8]
            Economics
            Astronomy - distribution of galaxies; weather patterns

      Origins of these phenomena

2. The Goals of a Theory of Complexity / Theories of Complexity
      Types of models -- ODE's, PDE's, etc.
      The importance of computers -- programs as models
      The periodic/fractal/chaotic sequence in terms of rewrite rules --
            the appearance of 1/f noise.
          - intermittency
          - Relation to generating functions; functional equations;
                  word problems for groups
      Basic principles; randomness doesn't make a fundamental difference

3. Cellular Automata -- basic phenomenology /
            The Phenomenology of Cellular Automata
      Statistical behavior
      1,2,3 dimensions
      Approximate conservation laws
      "Naturally-occurring rules" vs. specially constructed ones
      Periodic / fractal / chaotic

3a. Cellular automaton-based modelling / Modelling with Cellular Automata
      Reaction-diffusion systems
            Chemistry
            Biology
      Hydrodynamics
      Dendritic growth
      Spin systems
      Vision systems

3'. Global analysis of cellular automata / Global Analyses of Cellular Automata
      Attractors
      Entropy etc.
      Linear CA

4. Complexity in numbers -- traditional dynamical systems theory /
            Numerical Models of Complex Systems
      Iterated maps
      Differential equations
      Number theory
```
```
            of models
               s
               systems
               ters
         Critical phenomena; the renormalization group
         inear; solitons; non-linear
         etworks
         ed growth models -
            - percolation
            - Eden model
            - DLA
         l mechanics; 3-body problem etc.
         olean networks

         heory
         f computation
            Serial
            Parallel
         Constructing computations in CA
         Computational spaces (e.g. formal languages)
            Characterizing CA sets
         Classes of computations
         Intractability
            Approximate solutions
               Simulated annealed
               Hypermetricity
         Undecidability

5a. Computational irreducibility and its consequences
         Applications of undecidability to physics
         Applications of intractability
         The need for simulation

5'. Data analysis etc.
         The idea of statistics
         Finding parameters of models
         The Fourier transform & fractal transform
         Perceptrons & other models of human perception -- our main
             analyzer of complex systems

6. The nature of randomness
         Probabilistic models - why almost any random generator works
         Noise as a way of representing hidden degrees of freedom
         The dithering problem -- approximating continuous systems
         Cryptography
         Quantum mechanics

7. Thermodynamics & large-scale limits
         Reversibility & dissipation
         The arrow of time
         Fluctuations
         Lattice independence; e.g. in Eden, DLA, random walk, CA

9. Defining complexity

10. Systems that adapt
         Models of self-reproduction
         'Developmental constraints'
         Results from core wars
         Trends in trilobites and other things (e.g. programs etc.)
```

I wrote a few pages of introductory text—beginning by stating the objective as:

```
The purpose of this book is to develop a science of complexity.
Our goal is to find general principles that allow us to describe and
explain the many instances of complexity in the natural and
artificial world.
```

My archives record that in late December I was taking a more computation-first approach, and considering the title *Algorithms in Nature: An Introduction to Complexity*. But soon I was submerged in the intense effort to develop Mathematica 2.0, and this is what consumed me for most of 1990—though my archives from the time reveal one solitary short note, apparently from the middle of the year:

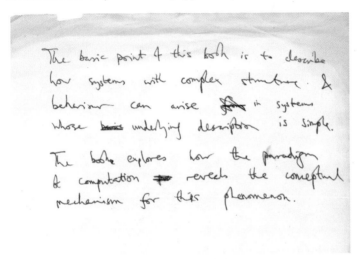

But through all this I kept thinking about the book I intended to write, and wondering what it should really be like. In the late 1980s there'd been quite a run of unexpectedly successful "popular science" books—like *A Brief History of Time*—that mixed what were at least often claimed to be new results or new insights about science with a kind of intended-to-entertain "everyman narrative". A sequence of publishers had encouraged me to "write a popular science book". But should the book I was planning to write really be one of those?

I talked to quite a few authors and editors. But nobody could quite tell a coherent story. Perhaps the most promising insight came from an editor of several successful such books, who opined that he thought the main market for "popular science" books was people who in the past would have read philosophy books, but now those were too narrow and technical. Other people, though, told me they thought it was really more of an "internal market", with the books basically being bought by other scientists. And in the media and elsewhere there continued to be an undercurrent of sentiment that while the books might be being bought, they mostly weren't actually getting read.

"Isn't there actual data on what's going on?" I asked my publishing industry contacts. "No", they said, "that's just not how our industry works". "Well", I said, "why don't we collect some data?" My then-publisher seemed enthusiastic about it. So I wrote a rather extensive survey to do on "random shoppers" in bookstores. It began with some basic—if "1990-style"—demographic questions, then got to things like

**23. Which of the following scientists and others have you heard of?**

- ☐ John Conway
- ☐ Francis Crick
- ☐ Freeman Dyson
- ☐ Mitchell Feigenbaum
- ☐ Richard Feynman
- ☐ Martin Gardner
- ☐ Bill Gates
- ☐ Murray Gell-Mann
- ☐ Stephen Hawking
- ☐ Danny Hillis
- ☐ Steven Jobs
- ☐ Stuart Kauffman
- ☐ Donald Knuth
- ☐ Chris Langton
- ☐ Benoit Mandelbrot
- ☐ Marvin Minsky
- ☐ Carl Sagan
- ☐ Terry Sejnowski
- ☐ Claude Shannon
- ☐ Larry Smarr
- ☐ Bill Thurston
- ☐ Alan Turing
- ☐ John von Neumann
- ☐ Stephen Wolfram

**24. For which of these books have you done the following?**

| | Bought and read | Bought and not read | Looked at | Read reviews | Heard about | Never heard of |
|---|---|---|---|---|---|---|
| Albert Einstein: *The Meaning of Relativity* | ☐ | ☐ | ☐ | ☐ | ☐ | ☐ |
| Daniel Dennett: *Consciousness Explained* | ☐ | ☐ | ☐ | ☐ | ☐ | ☐ |
| Douglas Hofstadter: *Gödel, Escher, Bach* | ☐ | ☐ | ☐ | ☐ | ☐ | ☐ |
| Fritzof Capra: *Tao of Physics* | ☐ | ☐ | ☐ | ☐ | ☐ | ☐ |
| James Gleick: *Chaos* | ☐ | ☐ | ☐ | ☐ | ☐ | ☐ |
| Leighton/Feynman: *Surely You're Joking, Mr. Feynman* | ☐ | ☐ | ☐ | ☐ | ☐ | ☐ |
| Peitgen/Richter: *Beauty of Fractals* | ☐ | ☐ | ☐ | ☐ | ☐ | ☐ |
| Roger Penrose: *Emperor's New Mind* | ☐ | ☐ | ☐ | ☐ | ☐ | ☐ |
| Stephen J. Gould: *Wonderful Life* | ☐ | ☐ | ☐ | ☐ | ☐ | ☐ |
| Steven Levy: *Artificial Life* | ☐ | ☐ | ☐ | ☐ | ☐ | ☐ |
| Steven Weinberg: *First Three Minutes* | ☐ | ☐ | ☐ | ☐ | ☐ | ☐ |

**25. What do you think of the following fields?**

| | Very interesting | Fairly interesting | Old hat | Waste of time | Never heard of it |
|---|---|---|---|---|---|
| Artificial intelligence | ☐ | ☐ | ☐ | ☐ | ☐ |
| Artificial life | ☐ | ☐ | ☐ | ☐ | ☐ |
| Catastrophe theory | ☐ | ☐ | ☐ | ☐ | ☐ |
| Chaos theory | ☐ | ☐ | ☐ | ☐ | ☐ |
| Complexity theory | ☐ | ☐ | ☐ | ☐ | ☐ |
| Cosmology | ☐ | ☐ | ☐ | ☐ | ☐ |
| Paleontology | ☐ | ☐ | ☐ | ☐ | ☐ |
| Extraterrestrial intelligence | ☐ | ☐ | ☐ | ☐ | ☐ |
| Genetic engineering | ☐ | ☐ | ☐ | ☐ | ☐ |
| Neural networks | ☐ | ☐ | ☐ | ☐ | ☐ |
| Planetary exploration | ☐ | ☐ | ☐ | ☐ | ☐ |
| Quantum theory | ☐ | ☐ | ☐ | ☐ | ☐ |
| Unified field theories | ☐ | ☐ | ☐ | ☐ | ☐ |

**26. Which of the following have you heard of?**

- ☐ Black holes
- ☐ Bucky balls
- ☐ Cellular automata
- ☐ The game of life
- ☐ Introns
- ☐ The Jahn-Teller effect
- ☐ *Mathematica*
- ☐ NP completeness
- ☐ Polywater
- ☐ Pulsars
- ☐ Quarks
- ☐ RISC processors
- ☐ Turing machines
- ☐ The Z boson

and rather charmingly ended with

> **31. Do you use a computer?**
> ☐ At your office/school, every day
> ☐ At your office/school, sometimes
> ☐ At home, every day
> ☐ At home, sometimes
> ☐ Not often
>
> **32. Do you use electronic mail/on-line bulletin boards?**
> ☐ Not at all
> ☐ Internet
> ☐ CompuServe
> ☐ Prodigy
> ☐ Internal network within my organization
> ☐ Other services
>
> **33. When do you think electronic books will become common?**
> ☐ Here today
> ☐ Within 2 years
> ☐ Within 5 years
> ☐ Within 10 years
> ☐ Within 20 years
> ☐ Never

(and, yes, in reality it took almost the longest time I could imagine for electronic books to become common). But after many months of "we'll get results soon" it turned out almost no surveys were ever done. As I would learn repeatedly, most publishers seemed to have a very hard time doing anything they hadn't already done before. Still, my then-publisher had done well with *The Mathematica Book*. So perhaps they might be able to just "follow a formula" and do well with my book if it was written in "popular science" form.

But I quickly realized that the pressure to add sensationalism "to sell books" really grated on me. And it didn't take long to decide that, no, I wasn't going to write a "formula" popular science book. I was going to write my own kind of book—that was more direct and straightforward. No stories. Just science. With lots of pictures. And if nothing else, the book would at least be helpful to me, as a way of clarifying my own thinking.

## Beginning to Tool Up

In January 1991 we announced Mathematica 2.0—and in March and June I did a 35-city tour of the US and Europe talking about it. Then, finally, at the beginning of July we delivered final floppy disks to the duplicator (as one did in those days)—and

Mathematica 2.0 was on its way. So what next? I had a long roadmap of things we should do. But I decided it was time to let the team I'd built just get on with following the roadmap for a while, without me adding yet more things to it. (As it turns out, we finally finished essentially everything that was on my 1991 to-do list just a few years ago.)

And so it was that in July 1991 I became a remote CEO (yes, a few decades ahead of the times), moved a couple thousand miles away from our company headquarters to a place in the hills near San Francisco, and set about getting ready to write. Based on the plan I had for the book—and my experience with *The Mathematica Book*—I figured it might take about a year, or maybe 18 months, to finish the project.

In the end—with a few trips in the middle, notably to see a total solar eclipse—it took me a couple of months to get my remote-CEO setup figured out (with a swank computer-connected fax machine, email getting autodelivered every 15 minutes, etc.). But even while that was going on, I was tooling up to get an efficient modern system for visualizing and studying cellular automata. Back when I had been writing my papers in the 1980s, I'd had a C program (primarily for Sun workstations) that had gradually grown, and was eventually controlled by a rather elaborate—but sensible-for-its-time—hierarchical textual menu system

```
|////////////////////////////////////////////////////////|
|                                                        |
|           Cellular Automaton Simulation Package        |
|                                                        |
|////////////////////////////////////////////////////////|

Basic options:

Generate 1D cellular automaton patterns:
        1  Elementary rules (k=2, r=1)
        2  Additive (linear) rules
        3  Totalistic rules
        4  Functional (large k) rules
        5  Multiple lattice rules
        6  General rules
        7  Reversible (second-order) rules
        8  Continuous dynamics

Other options:
        9  Use parameter file
       10  Global analysis (structure search, entropies, languages, etc)
       11  Local analysis (Lyapunov exponents, Cantor sets, etc)
       12  Run selection dynamics
       13  Exit

Enter option : 10

        1  Generate finite cellular automaton state transition diagram.
        2  Find predecessors for spatial sequences in infinite cellular automata.
        3  Find predecessors for temporal sequences in infinite cellular automata.
        4  Search for persistent configurations.
        5  Monte Carlo search for cycles.
        6  Test for global surjectiveness (old version).
        7  Construct finite state machine configuration recognizer.
        8  Find excluded blocks.
        9  Sinai measure entropy algorithm
```

which, yes, could generate at least single-graphic-per-screen graphics, as in this picture of my 1983 office setup:

But now the world had changed, and I had Mathematica. And I wanted a nice collection of Wolfram Language functions that could be used as streamlined "primitives" for studying cellular automata. Given all my work on cellular automata it might seem strange that I hadn't built cellular automaton functionality into the Wolfram Language right from the start. But in addition to being a bit bashful about my personal pet kind of system, I hadn't been able to see how to "package" all the various different kinds of cellular automata I'd studied into one convenient superfunction—and indeed it took me a decade more of understanding, both of language design and of cellular automata, to work out how to nicely do that. And so back in 1991 I just created a collection of add-on functions (or what might today be a paclet) containing the particular functions I needed. And indeed those functions served me well over the course of the development of *A New Kind of Science*.

A "staged" screen capture from the time shows my basic working environment:

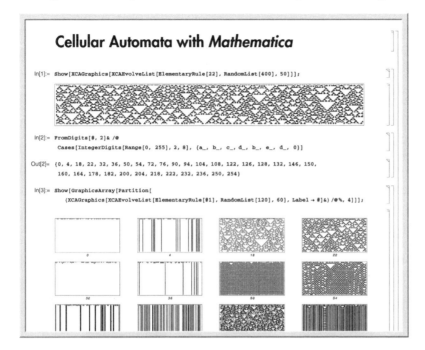

Some printouts from early 1991 give a sense of my everyday experience:

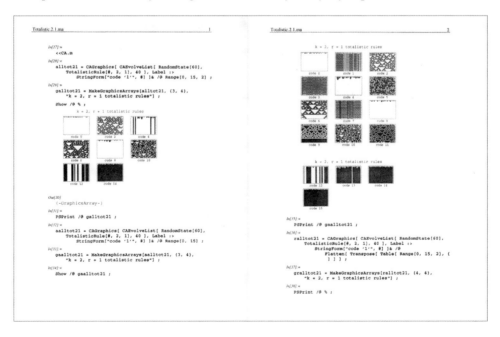

And although it's now more than 30 years later, I'm happy to say that we've successfully maintained the compatibility of the Wolfram Language, and those same functions still just run! The .ma format of my Version 2.0 notebooks from 1991 has to be converted to .nb, but then they just open in Version 13 (with a bit of automatic style modernization) and I'm immediately "transported back in time" to 1991, with, yes, a very small notebook appropriate for a 1991 rather than a 2022 screen size:

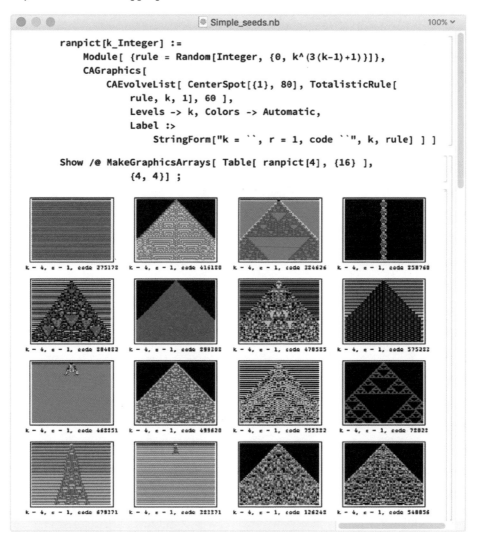

(Of course the cellular automata all look the same, but, yes, this notebook looks shockingly similar to ones from our recent cellular automaton NFT-minting event.)

We'd invented notebooks in 1987 to be able to do just the kinds of things I wanted to do for my science project—and I'd been itching to use them. But before 1991 I'd mostly been doing core code development (often in C), or using the elaborate but still textual system we had for authoring *The Mathematica Book*. And so—even though I'd demoed them many times—I hadn't had a chance to personally make daily use of notebooks.

But in 1991, I went all in on notebooks—and have never looked back. When I first started studying cellular automata back in 1981, I'd had to display their output as text. But soon I was able to start using the bitmapped displays of workstation computers, and by 1984 I was routinely printing cellular automaton images in fairly high resolution on a laser printer. But with Mathematica and our notebook technology things got dramatically more convenient—and what had previously often involved laborious work with paper, scissors and tape now became a matter of simple Wolfram Language code in a notebook.

For almost a decade starting in 1982, my primary computer had been a progressively more sophisticated Sun workstation. But in 1991 I switched to NeXT—mainly to be able to use our notebook interface, which was by then well developed for NeXT but wasn't yet ready on X Windows and Sun. (It was also available on Macintosh computers, but at the time those weren't powerful enough.)

And here I am in 1991, captured "hiding out" as a remote CEO, with a NeXT in the background, just getting started on the book:

Here's a picture showing a bit more of the setup, taken in early 1993, during a short period when I was a remote-remote-CEO, with my computer set up in a hotel room:

## September 1991: Beyond Cellular Automata

Throughout the 1980s, I'd used cellular automata—and basically cellular automata alone—as my window into the computational universe. But in August 1991—with my new computational capabilities and new away-from-the-company-to-do-science setup—I decided it'd be worth trying to look at some other systems.

And I have to say that now, three decades later, I didn't remember just how suddenly everything happened. But my filesystem records that in successive days at the beginning of September 1991 there I was investigating more and more kinds of systems (.ma's were "Mathematica notebook" files; .mb's were the "binary forks" of these files):

```
      549 Jan   5  1991 fourier.m
      170 Feb   3  1991 hof.m
  1379053 Sep   1  1991 MobileAutomata.ma
    14266 Sep   2  1991 MA-machines.mb.orig
    13846 Sep   2  1991 MA-machines.ma
     3887 Sep   2  1991 MA-beaver.ma
   250630 Sep   2  1991 tmp.ma
   606559 Sep   3  1991 TagSystems.ma
    52331 Sep   3  1991 turing.network
  2066530 Sep   3  1991 DMA-k=2.ma
     3713 Sep   3  1991 MA.m
  1003505 Sep   4  1991 Turing-random.ma
  1680092 Sep   4  1991 DMA-paths.ma
      524 Sep   4  1991 Tag.m
   259368 Sep   4  1991 TagSystems2.ma
     1672 Sep   5  1991 tagtime.tm.c
   106563 Sep   5  1991 tagtime
  3231916 Sep   6  1991 TagTimes.ma
  3088422 Sep  10  1991 Turing-2.ma
     3900 Feb  16  1992 turing.m
  7226381 Feb  17  1992 TuringNew.ma
 18500124 Feb  17  1992 TuringBigResults.nb
  8566368 Feb  17  1992 TuringBigResults.ma
```

Mobile automata. Turing machines. Tag systems. Soon these would be joined by register machines, and more. The first examples of these systems tended to have quite simple behavior. But I quickly started searching to see whether these systems—like cellular automata—would be capable of complex behavior, as my 1991 notebooks record:

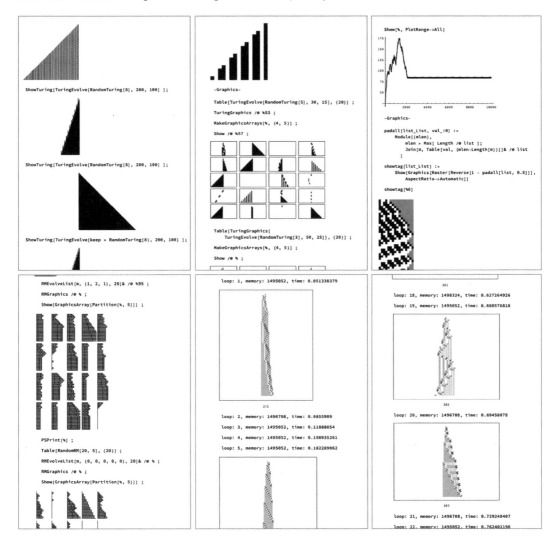

Often I would run programs overnight, or sometimes for many days. Later I would recruit many computers from around our company, and have them send me mail about their results:

```
Date: Mon, 6 Sep 1993 23:18:07 -0600
From: swolf@wri.com
Apparently-To: swolf@wri.com

~s F14 passed - 11592 23374

Host: pyrethrum; PID: 14307

Stats: {tot: 156290, F4: 16, P5: 151120, F6: 3029, F8: 1526, P8: 410, F10: 122, P10: 42, S: 25}

Rule: 11592 23374

Result: F14 passed

{
{0, 0, 0, 0, 0, 1, 1, 1, 0, 0, 0, 0, 0, 0},
{0, 0, 1, 0, 0, 1, 1, 0, 0, 1, 0, 0, 1, 0},
{0, 0, 1, 1, 1, 1, 0, 0, 0, 1, 1, 1, 1, 0},
{0, 0, 0, 1, 1, 0, 0, 1, 0, 0, 1, 1, 0, 0},
{1, 1, 1, 1, 0, 0, 0, 1, 1, 1, 1, 0, 0, 0},
```

But already in September 1991 I was starting to see that, yes, just like cellular automata, all these different kinds of systems, even when their underlying rules were simple, could exhibit highly complex behavior. I think I'd sort of implicitly assumed this would be true. But somehow actually seeing it began to elevate my view of just how general a "science of complexity" one might be able to make.

There were a few distractions in the fall of 1991. Like in October a large fire came within about half a mile of burning down our house:

But by the spring of 1992 it was beginning to become clear that there was a very general principle around all this complexity I was seeing. I had invented the concept of computational irreducibility back in 1984. And I suppose in retrospect I should have seen the bigger picture sooner. But as it was, on a pleasant afternoon (and, no, I haven't figured out the exact date), I was taking a short break from being in front of my computer, and had wandered outside. And that's when the Principle of Computational Equivalence came to me. Somehow after all those years with cellular automata, and all those months with computer experiments on other systems, I was primed for it. But in the end it all arrived in one moment: the concept, the name, the implications for computational irreducibility. And in the three decades since, it's been the single most important guiding principle for my intuition.

## What Should the Pages Look Like?

I've always found it difficult to produce "disembodied content": right from the beginning I typically need to have a pretty clear idea how what I'm producing will look in the end. So back in 1991 I really couldn't produce more than a page or two of content for my book without knowing what the book was going to look like.

"Formula" popular science books tended—for what I later realized were largely economic reasons—to consist mainly of pages of pure text, with at most line drawings, and to concentrate whatever things like photographs they might have into a special collection of "plates" in the middle of the book. For *The Mathematica Book* we'd developed a definite—very functional—layout, with text, tables and two-column "computer dialogs":

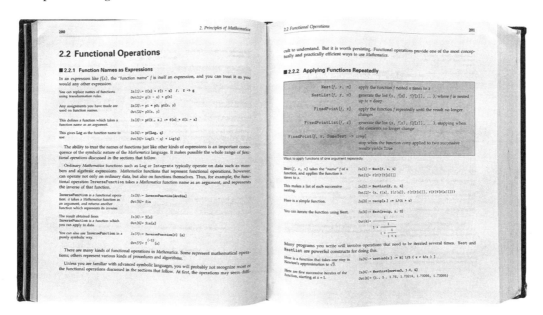

For the NKS book I knew I needed something much more visual. And at first I imagined it might be a bit like a high-end textbook, complete with all sorts of structured elements ("Historical Note", "Methodology", etc.).

I asked a talented young designer who had worked on *The Mathematica Book* (and who, 31 years later, is now a very senior executive at our company) to see what he could come up with. And here, from November 1991, is the very first "look" for the NKS book—with content pretty much just flowed in from the few pages I'd written out in plain text:

I knew the book would have images of the kind I'd long produced of cellular automata, and that had appeared in my papers and book from the 1980s:

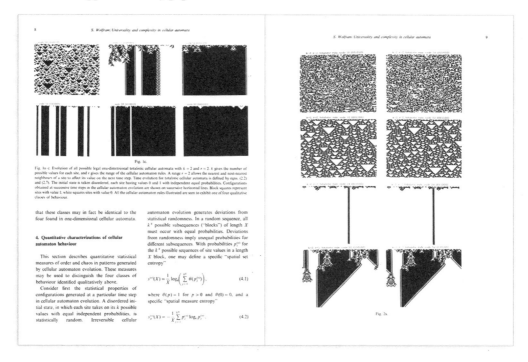

But what about "diagrams"? At first we toyed with drawing "textbook-style" diagrams—and produced some samples:

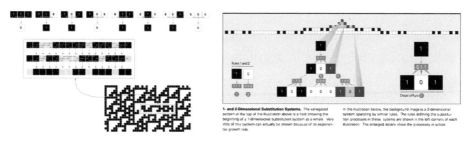

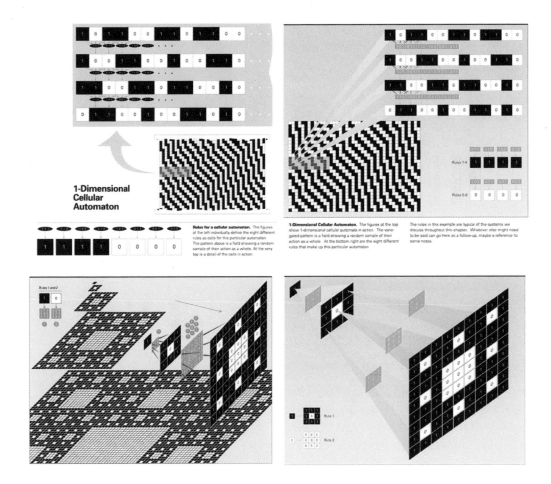

But these seemed to have way too much "conceptual baggage", and when one looks closely at them, it's easy to get confused. I wanted something more minimal—where the spotlight was as much as possible on the systems I was studying, not on "diagrammatic scaffolding". And so I tried to develop a "direct diagramming" methodology, where each diagram could directly "explain itself"—and where every diagram would be readable "purely visually", without words.

In a typical case I might show the behavior of a system (here a mobile automaton), next to an explicit "visual template" of how its rules operate. The idea then was that even a reader who didn't understand the bigger story, or any of the technical details, could still "match up templates" and understand what was going on in a particular picture:

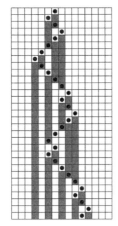

An example of a mobile automaton. Like a cellular automaton, a mobile automaton consists of a line of cells, with each cell having two possible colors. But unlike a cellular automaton, a mobile automaton has only one "active cell" (indicated here by a black dot) at any particular step. The rule for the mobile automaton specifies both how the color of this active cell should be updated, and whether it should move to the left or right. The result of evolution for a larger number of steps with the particular rule shown here is given as example (f) on the next page.

At the beginning of the project, the diagrams were comparatively simple. But as the project progressed I invented more and more mechanisms for them, until later in the project I was producing very complex "visually readable" diagrams like this:

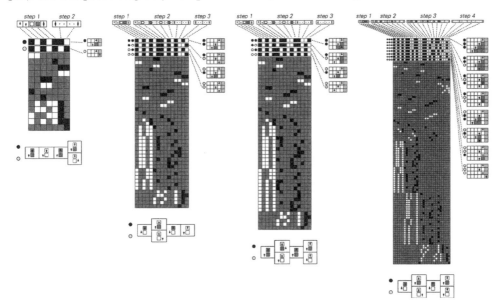

A crucial point was that all these diagrams were being produced algorithmically—with Wolfram Language code. And in fact I was developing the diagrams as an integral part of actually doing the research for the book. It was a lesson I'd learned years earlier: don't wait until research is "finished" to figure out how to present it; work out the presentation as early as possible, so you can use it to help you actually do the research.

Another aspect of our first "textbook-like" style for the book was the idea of having additional elements, alongside the "main narrative" of the book. In early layouts we thought about having "Technical Notes", "Historical Notes", "Implementation Notes", etc. But it didn't take too long to decide that no, that was just going to be too complicated. So we made the decision to have one kind of note, and to collect all notes at the back of the book.

And that meant that in the main part of the book we had just two basic elements: text and images (with captions). But, OK, in designing any book a very basic question is: what size and shape will its pages be? *The Mathematica Book* was squarish—like a typical textbook—so that it accommodated its text-on-the-left code-on-the-right "dialogs". We knew that the new book should be wide too, to accommodate the kinds of graphics I expected. But that posed a problem.

In *The Mathematica Book* ordinary text ran the full width of the page. And that worked OK, because in that book the text was typically broken up by dialogs, tables, etc. In the new book, however, I expected much longer blocks of pure text—which wouldn't be readable if they ran the full width of the page. But if the text was narrower, then how would the graphics not look like they were awkwardly sticking out? Well, the pages would have to be carefully laid out to appropriately anchor the graphics visually, say to the tops or bottoms of pages. And that was going to make the process of layout much trickier.

Different pages were definitely going to look different. But there had to be a certain overall consistency. Every graphic was going to have a caption—and actually a caption that was sufficiently self-contained so that people could basically "read the book just by looking at the pictures". Within the graphics themselves there had to be standards. How should arrays of cells be rendered? To what extent should things have boxes around them, or arrows between them? How big should pictures that emphasized particular features be?

Some of these standards got implemented basically just by me remembering to follow them. But others were essentially the result of the whole stack of Wolfram Language functions that we built to produce the algorithmic diagrams for the book. At the time, there was some fiddliness to these functions, and to making their output look good—though in later years what we learned from this was used to tune up the general look of built-in graphics in the Wolfram Language.

## The Technology of Images

One of the striking features of the NKS book is the crispness of its pictures. And I think it's fair to say that this wasn't easy to achieve—and in the end required a pretty deep dive into the technology of imaging and printing (as I'll describe more in a later section).

Back in the 1980s I'd had plenty of pictures of things like cellular automata in my papers. And I'd produced them by outputting what amounted to pages of bitmaps on laser printers, then having publishers photographically reproduce the pictures for printing.

Up to a point the results were OK:

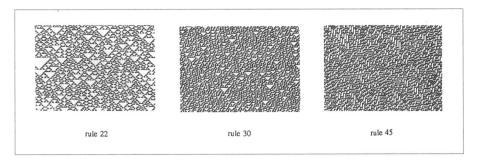

rule 22    rule 30    rule 45

But for example in 1985 when I wanted a 2000-step picture of rule 30 things got difficult. The computation (which, yes, involves 8 million cells) was done on a prototype Connection Machine parallel computer. And at first the output was generated on a large-format printer that was usually used to print integrated circuit layouts. The result was quite large, and I subsequently laminated pictures like this (and in rolled-up form they served as engaging hiding places for my children when they were very young):

But when photographically reproduced and printed in a journal the picture definitely wasn't great:

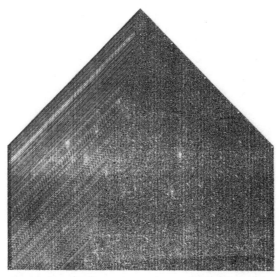

FIG. 6.1. Patterns generated by evolution for 250 and 2000 generations, respectively, according to the cellular automaton rule (3.1) from an initial state containing a single nonzero site. (The second pattern was obtained by Jim Salem using a prototype Connection Machine computer.)

And the NKS book provided another challenge as well. While the core of a picture might just be an array of cells like in a cellular automaton, a full algorithmic diagram could contain all sorts of other elements.

In the end, the NKS book was a beneficiary of an important design decision that we made back in 1987, early in the development of Mathematica. At the time, most graphics were thought about in terms of bitmaps. On whatever device one was using, there was an array of pixels of a certain resolution. And the focus was on rendering the graphics at that resolution. Not everything worked that way, though. And "drawing" (as opposed to "painting") programs typically created graphics in "vector" form, in which at first primitives like lines and polygons were specified without reference to resolution, and were then converted to bitmaps only when they were displayed.

The shapes of characters in fonts were something that was often specified—at least at an underlying level—in vector form. There'd been various approaches to doing this, but by 1987 PostScript was an emerging standard—at least for printing—buoyed by its use in the Apple LaserWriter. The main focus of PostScript was on fonts and text, but the PostScript language also included standard graphics primitives like lines and polygons.

Back when I had built SMP in 1979–1981 we'd basically had to build a separate driver for every different display or printing device we wanted to output graphics on. But in 1987 there was an alternative: just use PostScript for everything. Printer manufacturers were working hard to support PostScript on their printers, but PostScript mostly hadn't come to screens yet. There was an important exception though: the NeXT computer was set up to have PostScript as its native screen-rendering system. And partly through that, we decided to use PostScript as our underlying way to represent all graphics in Mathematica.

At a high level, graphics were described with the same symbolic primitives as we use in the Wolfram Language today: Line, Polygon, etc. But these were converted internally to PostScript—and even stored in notebooks that way. On the NeXT this was pretty much the end of the story, but on other systems we had to write our own interpreters for at least the subset of PostScript we were using.

Why was this important to the NKS book? Well, it meant that all graphics could be specified in a fundamentally resolution-independent way. In developing the graphics I could look at them in a notebook on a screen, or I could print them on a standard laser printer. But for the final book the exact same graphics could be printed at much higher resolution—and look much crisper.

At the time, the standard resolution of a computer screen was 72 dpi (dots per inch) and the resolution of a typical laser printer was 300 dpi. But the typical basic resolution of a book-printing pipeline was more like 2400 dpi. I'll talk later about the adventure of actually printing the NKS book. But the key point was that because Mathematica's graphics were fundamentally based on PostScript, they weren't tied to any particular resolution, so they could in principle make use of whatever resolution was available.

Needless to say, there were plenty of complicated issues. One had to do with indicating the cells in something like a cellular automaton. Here's a picture of the first few steps of rule 30, shown as a kind of "macro bitmap", with pure black and white cells:

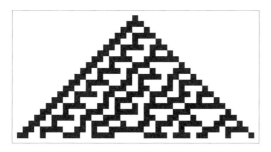

But often I wanted to indicate the extent of each cell:

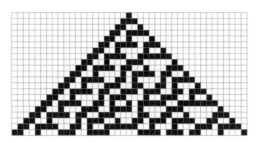

And in late 1991 and early 1992 we worried a lot about how to draw the "mesh" between cells. A first thought was just to use a thin black line. But that obviously wouldn't work, because it wouldn't separate black cells. And we soon settled on a GrayLevel[.15] line, which was visible against both black and white.

But how is such a line printed? If we're just using black ink, there's ultimately either black or white at a particular place on the page. But there's a standard way to achieve the appearance of gray, by changing the local density of black and white. And the typical method used to implement this is (as we'll discuss later) halftoning, in which one renders the "gray" by using black dots of different sizes.

But by the time one's using very thin gray lines, things are getting very tricky. For example, it matters how much the ink on either side of the line spreads—because if it's too much it can effectively fill in where the line was supposed to be. We wanted to define standards that we could use throughout the NKS book. And we couldn't tell what would happen in the final printed book except by actually trying it, on a real printing press. So already in early 1992 we started doing print tests, trying out different thicknesses of lines and so on. And that allowed us to start setting graphics standards that we could implement in the Wolfram Language code used to make the algorithmic diagrams, that would then flow through to all renderings of those diagrams.

Back in 1991 we debated quite a bit whether the NKS book should use color. We knew it would be significantly more expensive to print the book in color. But would color allow seriously better communication of information? Two-color cellular automata like rule 30 can be rendered in pure black and white. But over the years I'd certainly made many striking color pictures of cellular automata with more colors.

Somehow, though, those pictures hadn't seemed quite as crisp as the black and white ones. And there was another issue too, having to do with a problem I'd noticed in the mid-1980s in human visual perception of arrays of colored cells. Somewhat nerdily, I ended up including a note about this in the final NKS book:

> ■ **Using color.** Aside from practicalities of printing, what made me decide not to use color in this book were issues of visual perception. For much as it is easier to read text in black and white, so also it is easier to assimilate detailed pictures if they are just in black and white. And in fact many types of images in this book show quite misleading features in color. In human visual perception the color of something tends to seem different depending on what is around it—so that for example a red element tends to look purple or pink if the elements around it are respectively blue or white. And particularly if there are few colors arranged in ways that are not visually familiar it is typical for this effect to make all sorts of spurious patterns appear.

But the final conclusion was that, yes, the NKS book would be pure black and white. Nowadays—particularly with screen rendering being in many ways more important than print—it's much easier to do things in color. And, for example, in our Physics Project it's been very convenient to distinguish types of graphs, or nodes in graphs, by color. But for the NKS book I think it was absolutely the right decision to use black and white. Color might have added some nice accents to certain kinds of diagrams. But the clarity—and visual force—of the images in the book was much better served by the perceptual crispness of pure black and white.

## How to Lay Out the Book

The way most books with complex formats get produced is that first the author creates "disembodied" pieces of content, then a designer or production artist comes in and arranges them on pages. But for the NKS book I wanted something where the process of creation and layout was much more integrated, and where—just as I was directly writing Wolfram Language code to produce images—I could also directly lay out final book pages.

By 1990 "desktop publishing" was commonplace, and there were plenty of systems that basically allowed one to put anything anywhere on a page. But to make a whole book we knew we needed a more consistent and templated approach—that could also interact programmatically with the Wolfram Language. There were a few well-developed "full-scale book production systems" that existed, but they were complex "industrially oriented" pieces of software, that didn't seem realistic for me to use interactively while writing the book.

In mid-1990, though, we saw a demo of something new, running on the NeXT computer: a system called FrameMaker, which featured book-production capabilities, as well as a somewhat streamlined interchange format. Oh, and especially on the NeXT, it handled PostScript graphics well, inserting them "by reference" into documents. By late 1990 we were building book layout templates in FrameMaker, and we soon settled on using that for the basic production of the book. (Later, to achieve all the effects we wanted, we ended up having to process everything through Wolfram Language, but that's another story.)

We iterated for a while on the book design, but by the end of 1991 we'd nailed it down, and I started authoring the book. I made images using Mathematica, importing them in "Encapsulated PostScript" into FrameMaker. And words I typed directly into FrameMaker—in the environment reconstructed here using a virtual machine that we saved from the time of authoring the book:

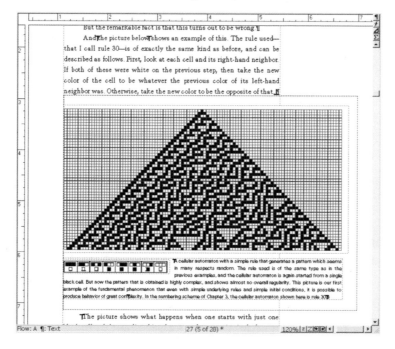

I composed every page—not only its content, but also its visual appearance. If I had a cellular automaton to render, and it was going to occupy a certain region on a page, I would pick the number of cells and steps to be appropriate for that region. I was constantly adjusting pictures to make them look good on a given page, or on pairs of facing pages, or along with other nearby pictures, and so on.

One of the tricky issues was how to refer to pictures from within the text. In technical books, it's common to number "figures", so that the text might say "See Figure 16". But I wanted to avoid that piece of "scaffolding", and instead always just be able to say things like "the picture below", or "the picture on the facing page". It was often quite a puzzle to see how to do this. If a picture was too big, or the text was too small, the picture would get too far ahead, and so on. And I was constantly adjusting things to make everything work.

I also decided that for elegance I wanted to avoid ever having to hyphenate words in the text. And quite often I found myself either rewording things, or slightly changing letter spacing, to make things fit, and to avoid things like "orphaned" words at the beginnings of lines.

It was a strange and painstaking process getting each page to look right, and adjusting content and layout together. Sometimes things got a little pathological. I always wanted to fill out pages, and not to leave space at the bottom (oh, and facing pages had to be exactly the same height). And I also tried to start new sections on a new page. But there I was, writing Chapter 5, and trying to end the section on "Substitution Systems and Fractals"—and I had an empty bottom third of a page. What was I to do? I decided to invent a whole new kind of system, that appears on page 192, just to fill out the layout for page 191:

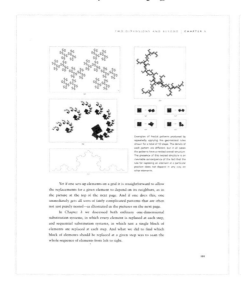 

Looking through my archives, I find traces of other examples. Here are notes on a printout of Chapter 6. And, yes, on page 228 I did insert images of additional rules:

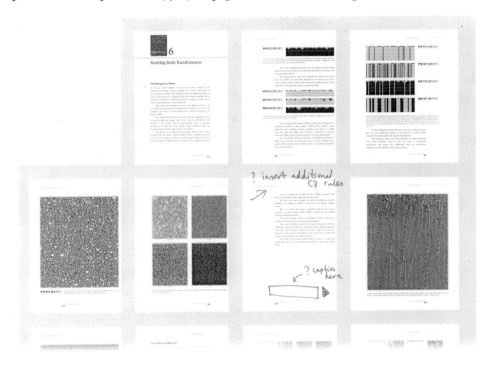

## The Book Takes Shape

By the end of 1991 I was all set up to author and lay out the book. I started writing—and things went quickly. The first printout I have from that time is from May 1992, and it already has nearly 90 pages of content, with many recognizable pictures from the final NKS book:

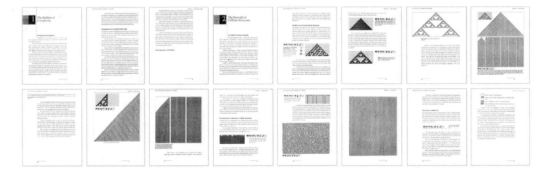

At that point the book was titled *Computation and the Complexity of Nature,* and the chapter titles were a bit different, and rather complexity themed:

**Contents**

Preface

1. The Problem of Complexity
2. The Example of Cellular Automata
3. Phenomena of Complexity
4. The Universality of Complexity
5. The Origin of Natural Forms
6. The Phenomenon of Computation
7. Computational Irreducibility
8. The Difficulty of Searching
9. The Origins of Randomness
10. Consequences of Complexity

General Notes.

Notes for all chapters.

Footnotes and References.

Index.

A large fraction of the main-text material about cellular automata was already there, as well as material about substitution systems and mobile automata. And there were extensive notes at the end, though at that point they were still single-column, and looked pretty much just like a slightly compressed version of the main text. And, by the way, Turing machines were just then appearing in the book, but still relegated to the notes, on the grounds that they "weren't as minimal as mobile automata".

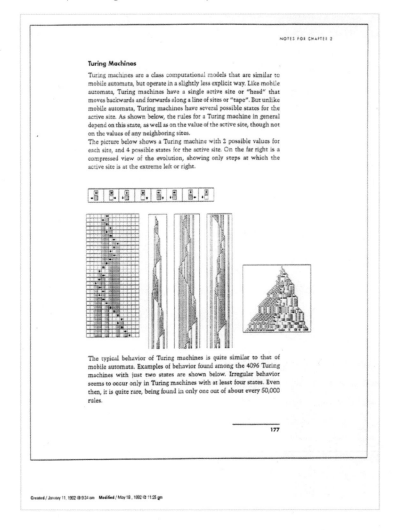

And hanging out, so far just as a stub, was the Principle of Computational Equivalence:

> **CHAPTER 1** / The Phenomenon of Computation
>
> by combining these structures in appropriate ways one can perform all the various operations that are for example implemented by individual sites with different values in the universal cellular automaton shown on page XXXX.
>
> This would have remarkable consequences. It would mean that within this one simple cellular automaton there is already all of the underlying structure needed to reproduce any kind of cellular automaton behavior, however complex it may be.
>
> **The Principle of Computational Equivalence**
>
> 11

By August 1992 the book had changed its title to *A New Science of Complexity* (subtitle: *Rethinking the Mechanisms of Nature*). There was a new first chapter "Some Fundamental Phenomena" that began with photographs of various "systems from nature":

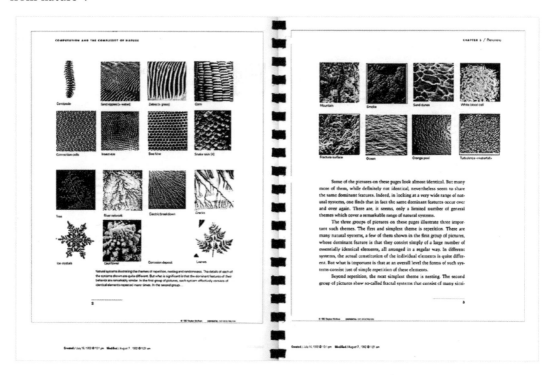

Chapter 3 had now become "The Behavior of Simple Systems". Turing machines were there. There was at least a stub for register machines and arithmetic systems. But even though I'd investigated tag systems in September 1991 they weren't yet in the book. Systems based on numbers were starting to be there.

And then, making their first appearance (with the page tagged as having been modified May 25, 1992), were the multiway systems that are now so central to the multi-computational paradigm (or, as I had originally and perhaps more correctly called them in this case, "Multiway Substitution Systems"):

**Multiway Substitution Systems**

All the sections we have discussed so far have the feature that each step in their evolution has one unique outcome. But it is also sometimes useful to consider systems in which each step can have several outcomes.

Multiway substitution systems are one example. In sequential substitution systems as discussed on page XXXX there are in general a set of possible replacement rules, but at each step only the first such rule that applies is ever used. In multiway substitution systems, however, all the rules that apply are used. At each step therefore each sequence in general gives rise to several new sequences.

One can represent the evolution of multiway substitution systems by networks in which each node corresponds to one sequence, just like in the state transition networks shown on page XXX. Every node in the network is joined to nodes which correspond to all the sequences that are obtained from it in one step of the evolution of the system.

In simple cases, like the one shown below, the network has a very regular form. [{{0}->{0, 0}, {0}->{0}}] <trees that do and do not have identical nodes combined>

In the simplest cases, only a limited number of sequences are ever generated, so the corresponding network is finite. In other cases, an infinite number of sequences can ultimately be generated, but the structure of the network is simple. This happens, for example, when each node branches in the same way at every step. The result is a tree-like network, analogous to that obtained from the neighbor-independent substitution systems discussed on page XXXXX.

29

By September 1992, register machines were in, complete with the simplest register machine with complex behavior (that had taken a lot of computer time to find). My simple PDE with complex behavior was also there. By early 1993 I had changed its name again, to *A Science of Complexity*, and had begun to have a quite recognizable chapter structure (though not yet with realistic page numbers):

## Contents

*Preface* vi

**PART 1: FUNDAMENTAL PHENOMENA**
*Chapter 1 / Complexity in Nature* 1
*Chapter 2 / The Behavior of Simple Programs* 17

**PART 2: DEVELOPING THE NEW INTUITION**
*Chapter 3 / A Diversity of Programs* 43
*Chapter 4 / Systems Based on Numbers* 57
*Chapter 5 / Two Dimensions and Beyond* 73
*Chapter 6 / The Problem of Satisfying Constraints* 89
*Chapter 7 / Efforts at Analysis* 103

**PART 3: THE MECHANISMS OF NATURE**
*Chapter 8 / Basic Issues* 123
*Chapter 9 / Origins of Everyday Phenomena* 167
*Chapter 10 / Processes of Perception* 167
*Chapter 11 / Issues of Fundamental Physics* 189

**PART 4: THE FRAMEWORK FOR A NEW SCIENCE**
*Chapter 12 / Behavior as Computation* 203
*Chapter 13 / The Principle of Computational Equivalence* 227
*Chapter 14 / Fundamental Implications* 245

It imagined a rather different configuration of notes than eventually emerged:

*Epilog* 277

**NOTES FOR SPECIALISTS**
*Notes for Pure Mathematicians* 000
*Notes for Applied Mathematicians* 000
*Notes for Statisticians* 000
*Notes for Computer Scientists* 000
*Notes for Physicists* 000
*Notes for Biologists* 000
*Notes for Economists* 000
*Notes for Philosophers* 000
*Notes for Complex Systems Researchers* 000
*Notes for Educators* 000

*Notes for Students* 000
*Notes on the Text* 000
*Acknowledgements* 000
*Index* 000

Making its first appearance was a chapter on physics, though still definitely as a stub:

This version of the book opened with "chapter summaries", noting about the chapter on fundamental physics that "[Its] high point is probably my (still speculative) attempt to reformulate the foundation of physics in computational terms, including new models for space, time and quantum mechanics":

> **Chapter 9: Everyday Phenomena**
>
> This chapter is a tour of a wide variety of physical, biological, geological, economic, and other phenomena. It discusses the shapes of snowflakes, the striping patterns on zebras, the shapes of coastlines, the fluctuations in stock markets, and many other topics. In each case, it discusses what is known about the overall behavior of the system, and about the mechanisms that produce it.
>
> **Chapter 10: Fundamental Physics**
>
> Physics has already succeeded in summarizing some aspects of our universe, so this chapter looks at how my ideas relate to physical laws as they are now known. The high point is probably my (still speculative) attempt to reformulate the foundations of physics in computational terms, including new models for space, time and quantum mechanics.
>
> **Chapter 11: Perception and Analysis**
>
> One of the central ideas of the book is to look at all phenomena in computational terms. This idea also extends to our processes of perception and analysis. This chapter investigates what we know about the processes involved in vision and other kinds of perception, and relates them to the computational paradigm described in the book.

By February 1994 I was getting bound mockups of the book made, with the final page size, though the wrong title and cover, and at that point only 458 pages (rather than the eventual 1280):

The two-column format for the notes at the back was established, and even though the content of notes for the still-complexity-themed first chapter were rather different from the way they ended up, some later notes already looked pretty much the same as they would in the final book:

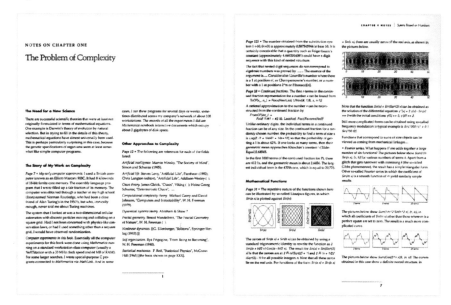

By September 1994 the draft of the book was up to 658 pages. The chapter structure was almost exactly as it finally ended up, albeit also with an epilog, and a bibliography (more about these later):

## Contents

| | | |
|---|---|---|
| *Preface* | | 7 |
| **PART 1: FUNDAMENTAL PHENOMENA** | | |
| *Chapter 1* / The Problem of Complexity | | 11 |
| *Chapter 2* / The Behavior of Simple Programs | | 33 |
| **PART 2: DEVELOPING THE NEW INTUITION** | | |
| *Chapter 3* / A Diversity of Programs | | 61 |
| *Chapter 4* / Systems Based on Numbers | | 127 |
| *Chapter 5* / Two Dimensions and Beyond | | 185 |
| *Chapter 6* / Starting from Randomness | | 233 |
| *Chapter 7* / Common Themes in Programs and Nature | | 313 |
| **PART 3: THE MECHANISMS OF NATURE** | | |
| *Chapter 8* / The Behavior of Everyday Systems | | 385 |
| *Chapter 9* / Fundamental Physics | | 415 |
| *Chapter 10* / Perception and Analysis | | 425 |
| **PART 4: THE FRAMEWORK FOR A NEW SCIENCE** | | |
| *Chapter 11* / The Notion of Computation | | 449 |
| *Chapter 12* / The Principle of Computational Equivalence | | 531 |
| *Epilog* / The Future of the Science in this Book | | 551 |
| CHAPTER NOTES | | 555 |
| BIBLIOGRAPHY | | 000 |
| INDEX | | 000 |

The September 1994 draft contained a section entitled "The Story of My Work on Complexity" (later renamed to the final "The Personal Story of the Science in this Book") which then included an image of what a Wolfram Notebook on NeXT looked like at the time:

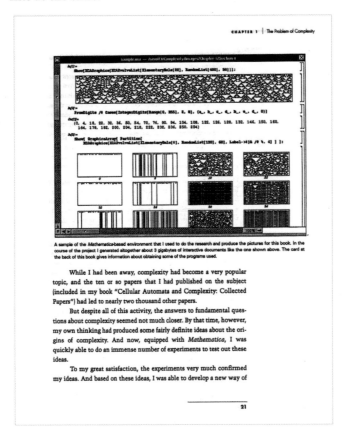

The caption talked about how in the course of the project I'd generated 3 gigabytes of notebooks—a number which would increase considerably before the book was finished. Charmingly, the caption also said: "The card at the back of this book gives information about obtaining some of the programs used". Our first corporate website went live on October 7, 1994.

By late 1994 the form of the book was basically all set. I'd successfully captured pretty much everything I'd known when I started on the book back in 1991, and I'd had three years of good discoveries. But what was still to come was seven years of intense research and writing that would take me much further than I had ever imagined back in 1991—and would end up roughly doubling the length of the book.

## Photographs for the Book

In 1991 I knew the book I was going to write would have lots of cellular automaton pictures. And I imagined that the main other type of pictures it would contain would be photographs of actual, natural systems. But where was I going to get those photographs from? There was no web with image search back then. We looked at stock photo catalogs, but somehow the kinds of images they had (often oriented towards advertising) were pretty far from what we wanted.

Over the years, I had collected—albeit a bit haphazardly—quite a few relevant images. But we needed many more. I wanted pictures illustrating both complexity, and simplicity. But the good news was that, as I explained early in the book, both are ubiquitous. So it should be easy to find examples of them—that one could go out and take nice, consistent photographs of.

And starting in late 1991, that's just what we did. My archives contain all sorts of negatives and contact prints (yes, this was before digital photography, and, yes, that's a bolt—intended as an example of simplicity in an artifact):

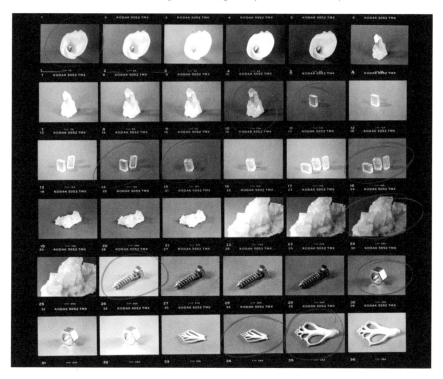

Sometimes the specimens I'd want could easily be found in my backyard

or in the sky

or on my desk (and even after waiting 400 million years, the trilobite fossil didn't make it in):

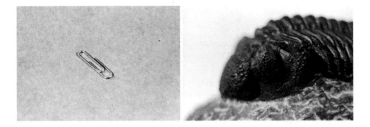

Over the course of a couple of years, I'd end up visiting all sorts of zoos, museums, labs, aquariums and botanical gardens—as well as taking trips to hardware stores and grocery stores—in search of interesting forms to photograph for the book.

Sometimes it would be a bit challenging to capture things in the field (yes, that's a big leaf I'm holding on the right):

At the zoo, a giraffe took a maddeningly long time to turn around and show me the other side of its patterning (I was very curious how similar they were):

There were efforts to get pictures of "simple forms" (yes, that's an egg)

with, I now notice, a cameo from me—captured in mid experiment:

Sometimes the subjects of photographs—with simple or complex forms—were acquired at local grocery stores (did I eat that cookie?):

I cast about far and wide for forms to photograph—including, I now realize, all of rock, paper and scissors, each illustrating something different:

Sometimes we tried to do actual, physical experiments, here with billiard balls (though in this case looking just like a simulation)

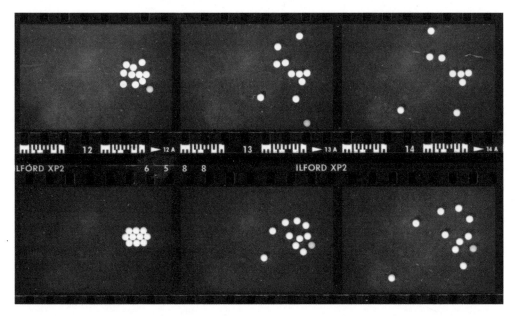

and here with splashes:

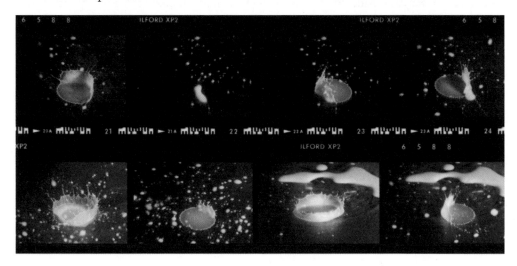

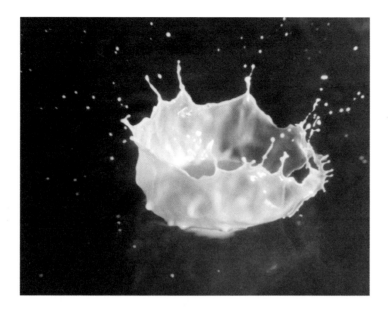

I was very interested in trying to illustrate reproducible apparently random behavior. I got a several-feet-tall piece of glassware at a surplus store and repeatedly tried dropping dye into water:

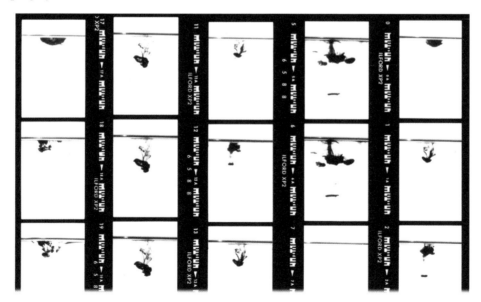

I tried looking at smoke rising:

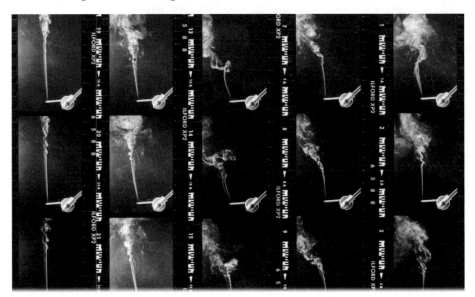

These were all do-it-yourself experiments. But that wasn't always enough. Here's a visit to a fluid dynamics lab (yes, with me visible checking out the hydraulic jump):

I'd simulated flow past an obstacle, but here it was "visualized" in real life:

Then there was the section on fracture. Again, I wanted to understand reproducibility. I got a pure silicon wafer from a physicist friend, then broke it:

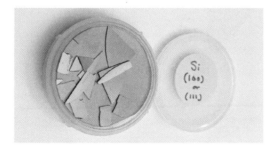

Under a powerful microscope, all sorts of interesting structure was visible on the fracture surface—that was useful for model building, even if not obviously reproducible:

And, talking of fractures, in March 1994 I managed to slip on some ice and break my ankle. Had I had pictures of fractures in the book, I was thinking of including an x-ray of my broken bones:

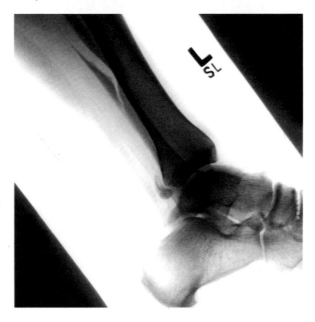

# The Making of *A New Kind of Science*

There are all sorts of stories about photographs that were taken for the book. In illustrating phyllotaxis (ultimately for Chapter 8), I wanted cabbage and broccoli. They were duly obtained from a grocery store, photographed, then eaten by the photographer (who reported that the immortalized cabbage was particularly tasty):

Another thing I studied in the book was shapes of leaves. Back in 1992 I'd picked up some neighborhood leaves where I was living in California at the time, then done a field trip to a nearby botanical garden. A couple of years later—believing the completion of the book was imminent—I was urgently trying to fill out more entries in a big array of leaf pictures. But I was in the Chicago area, and it was the middle of the winter, with no local leaves to be found. What was I to do? I contacted an employee of ours in Australia. Conveniently it turned out he lived just down the street from the Melbourne botanical gardens. And there he found all sorts of interesting leaves—making my final page a curious mixture of Californian and Australian fauna:

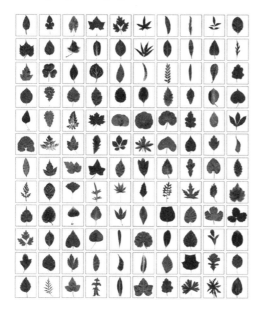

As it turned out, by the next spring I hadn't yet finished the book, and in fact I was still trying to fill in some of what I wanted to say about leaves. I had a model for leaf growth, but I wanted to validate it by seeing how leaves actually grow. That turned out not to be so easy—though I did dissect many leaf buds in the process. (And it was very convenient that this was a plant-related question, because I'm horribly squeamish when it comes to dissecting animals, even for food.)

Some of what I wanted to photograph was out in the world. But some was also collectible. Ever since I was a kid I had been gradually acquiring interesting shells, fossils, rocks and so on, sometimes "out in the field", but more often at shops. Working on the NKS book I dramatically accelerated that process. Shells were a particular focus, and I soon got to the point where I had specimens of most of the general kinds with "interesting forms". But there were still plenty of adventures—like finding my very best sample of "cellular-automaton-like" patterning, on a false melon volute shell tucked away at the back of a store in Florida:

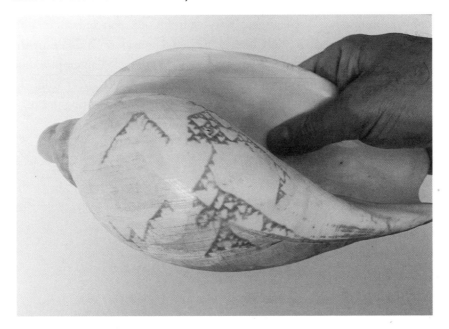

In 1998 I was working on the section of the book about biological growth, and wanted to understand the space of shell shapes. I was living in the Chicago area at that time, and spent a lovely afternoon with the curator of molluscs at the Field Museum of Natural History—gradually trying to fill in (with a story for every mollusc!) what became the array on page 416 of the book:

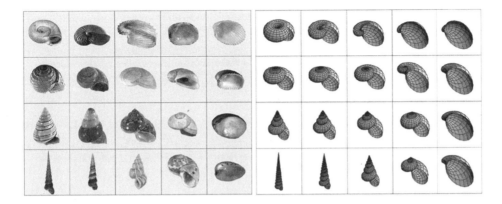

And actually it turned out that my own shell collection (with one exception, later remedied) already contained all the necessary species—and in a drawer in my office I still have the particular shells that were immortalized on that page:

I started to do the same kind of shape analysis for leaves—but never finished it, and it remains an open project even now:

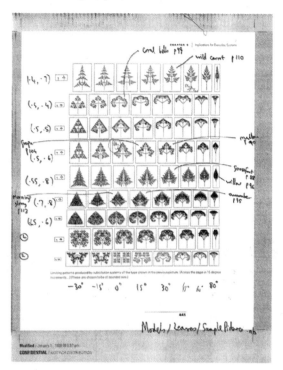

My original conception had been to start the book with "things we see in nature and elsewhere" and then work towards models and ideas of computation. But when I switched to "computation first" I briefly considered going to more "abstracted photographs", for example by stippling:

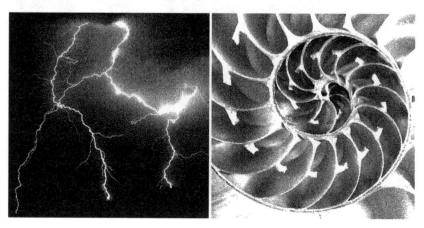

But in the end I decided that—just like my images of computational systems—any photographs should be as "direct as possible". And they wouldn't be at the beginning of the book, but instead would be concentrated in a specific later chapter (Chapter 8: "Implications for Everyday Systems"). Pictures of things like bolts and scissors became irrelevant, but by then I'd accumulated quite a library of images to choose from:

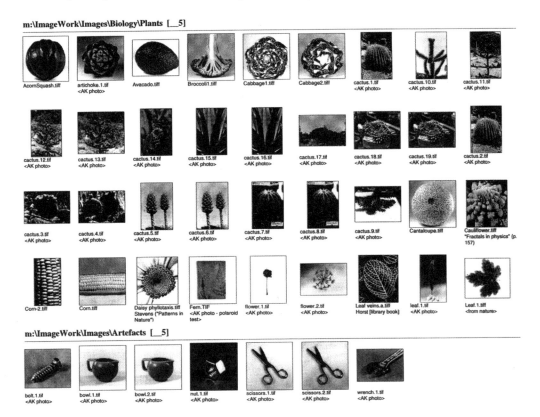

Many of these images did get used, but there were some nice collections, that never made it into the book because I decided to cut the sections that would discuss them. There were the "things that look similar" arrays:

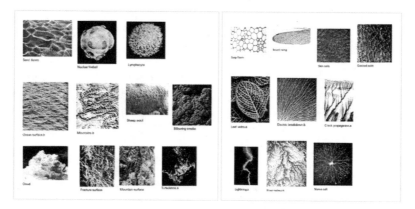

And there were things like pollen grains or mineral-related forms (and, yes, I personally crystallized that bismuth, which did at least make it into the notes):

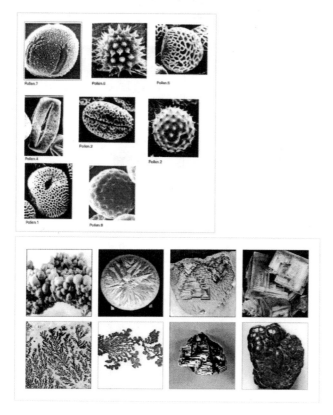

# The Making of A New Kind of Science

There were all sorts of unexpected challenges. I wanted an array of pictures of animals, to illustrate their range of pigmentation patterns. But so many of the pictures we could find (including ones I'd taken myself) we couldn't use—because I considered the facial expressions of the animals just too distracting.

And then there were stories like the "wild goose chase". I was sure I'd seen a picture of migrating birds (perhaps geese) in a nested, Sierpiński-like pattern. But try as we might, we couldn't find any trace of this.

But finally I began to assemble pictures into the arrays we were going to use. In the end, only a tiny fraction of the "nature" pictures we had made it into the book (and, for example, neither the egg nor the phyllotactically scaled pangolin here did)—some because they didn't seem clear in what they were illustrating, and some because they just didn't fit in with the final narrative:

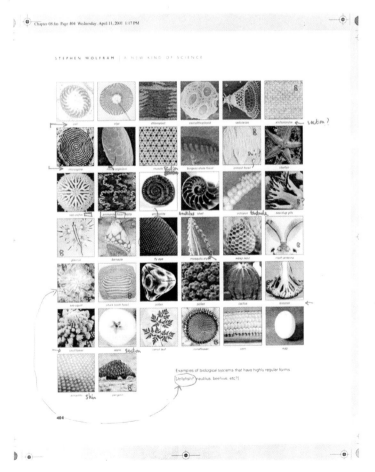

Beyond the natural world, the more I explored simple programs and what they can do, the more I wondered why so many of the remarkable things I was discovering hadn't been discovered before. And as part of that, I was curious what kinds of patterns people had in fact constructed from rules, for art or otherwise. On a few occasions during the time I was working on the book, I managed to visit relevant museums, searching for unexpected patterns made by rules:

But mostly all I could do was scour books on art history (and architecture) looking for relevant pictures (and, yes, it was books at the time—and in fact the web didn't immediately help even when it became available). Sometimes I would find a clear picture, and we would just ask for permission to reproduce it. But often I was interested in something that was for example off on the side in all the pictures we could find. So that meant we had to get our own pictures, and occasionally that was something of an adventure. Like when we got an employee of ours who happened to be vacationing in Italy to go to part of an obscure church in rural Italy—and get a photograph of a mosaic there from 1226 AD (and, yes, those are our photographer's feet):

### What Should the Book Be Called?

When I started working on the book in 1991 I saw it as an extension of what I'd done in the 1980s to establish a "science of complexity". So at first I simply called the book *The Science of Complexity*, adding the explanatory subtitle *A Unified Approach to Complex Behavior in Natural and Artificial Systems*. But after a while I began to feel that this sounded a bit stodgy—and like a textbook—so to spruce it up a bit I changed it to *A New Science of Complexity*, with subtitle *Rethinking the Mechanisms of Nature:*

A New Science of Complexity

Stephen Wolfram

Rethinking the Mechanisms of Nature

Pretty soon, though, I dropped the "New" as superfluous, and the title became *A Science of Complexity*. I always knew computation was a key part of the story, but as I began to understand more about just what was out there in the computational universe, I started thinking I should capture "computation" in the name of the book, leading to a new idea: *Computation and the Complexity of Nature*. And for this title I even had a first cover draft made—complete with an eye, added on the theory that human visual perception would draw people to the eye, and thus make them notice the book:

But back in 1992 (and I think it would be different today) people really didn't understand the term "computation", and it just made the book sound very technical to them. So back I went to *A Science of Complexity*. I wasn't very happy with it, though, and I kept on thinking about alternatives. In August 1992 I prepared a little survey:

```
Possible titles:

1. A Science of Complexity

2. Computation and the Complexity of Nature

3. From Simplicity to Complexity

4. A Science of Complex Things

5. The Origins of Complexity

6. On the Mechanisms of Nature

7. Complexity Explained


Subtitle:
How simple mechanisms explain complexity in nature and elsewhere


Rank these titles in terms of how:

A. Interesting the book would be to you

B. Understandable you would expect the book to be

C. Important the book is
```

The results of this survey were—like those of many surveys—inconclusive, and didn't change my mind about the title. Still, in October 1992 I dashed off an email considering *The Inevitable Complexity of Nature and Computation*. But 15 minutes later, as I put it, I'd "lost interest" in that, and it was back to *A Science of Complexity*.

By 1993, believing that the completion of the book was somehow imminent, we'd started trying to mock up the complete look of the book, including things like the back cover, and cover flaps:

The flap copy began: "This book is about a new kind of science that…". In the first chapter there was then a section called "The Need for a New Kind of Science":

As 1993 turned into 1994 I was still working with great intensity on the book, leaving almost no time to be out and about, talking about what I was doing. Occasionally, though, I would run into people and they would ask me what I was working on, and I would say it was a book, titled *A Science of Complexity*. And when I said that—at least among non-technical people—the reaction was essentially always the same "Oh, that sounds very complicated". And that would be the end of the conversation.

By September 1994 this had happened just too many times, and I realized I needed a new title. So I thought to myself "How would I describe the book?". And there it was, right in the flap copy: "a new kind of science". I made a quick note on the back of my then business card:

And soon that was the title: *A New Kind of Science*. I started trying it out. The reaction was again almost always the same. But now it was "So, what's new about it?" And that would start a conversation.

I liked the title a lot. It definitely said what by then I thought the book was about. But there was one thing I didn't like. It seemed a bit like a "meta title". OK, so you have a new kind of science. But what is that new kind of science called? What is its name? And why isn't the book called that?

I spent countless hours thinking about this. I thought about word roots. I considered comp- (for "computation"), prog- (for "program"), auto- (for "automata", etc.). I went through Latin and Greek dictionaries, and considered roots like arch- and log- (both way too confusing). I wrote programs to generate "synthetic words" that might evoke the right meaning. I considered names like "algonomics", "gramistry", "regulistics" (but not "ruliology"!), and "programistics"—for which I tried to see how its usage might work:

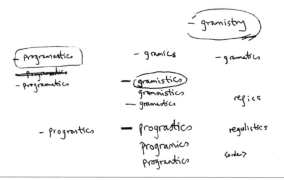

But nothing quite clicked. And in a sense my working title already told me why: I was talking about "a new kind of science", which involved a new way of thinking, for which there were really no words, because it hadn't been done before.

I'd had a certain amount of experience inventing words, for concepts in both science and technology. Sometimes it had gone well, sometimes not so well. And I knew the same was true in general in history. For every "physics" or "economics" or even "cybernetics" there were countless names that had never made it.

And eventually I decided that even if I could come up with a name, it wasn't worth the risk. Maybe a name would eventually emerge, and it would be perfectly OK if the "launch book" was called *A New Kind of Science* (as yet unnamed). Certainly much better than if it gave the new kind of science a definite name, but the name that stuck was different.

During the writing of *A New Kind of Science*, I didn't really need to "refer in the third person" to what the book was about. But pretty much as soon as the book was published, there needed to be a name for the intellectual endeavor that the book was about. During the development of the book, some of the people working on its project management had started calling the book by the initials of its title: ANKOS. And that was the seed for the name of its content, which almost immediately became "NKS".

Over the years, I've returned quite a few times to the question of naming. And very recently I've started using the term "ruliology" for one of the key pursuits of NKS: exploring the details of what systems based on simple computational rules do. I like the name, and I think it captures well the ethos of the specific scientific activity around studying the consequences of simple rules. But it's not the whole story of "NKS". *A New Kind of Science* is, as its name suggests, about a new kind of science—and a new way of thinking about the kind of thing we imagine science can be about.

When the book was first published, some people definitely seemed to feel that the strength and simplicity of the title "*A New Kind of Science*" must claim too much. But twenty years later, I think it's clear that the title said it right. And it's charming now when people talk about what's in *A New Kind of Science*, and how it's different from other things, and want to find a way to say what it is—and end up finding themselves saying it's "a new kind of science". And, yes, that's why I called the book that!

## The Cover of the Book

We started thinking about the cover of the book very early in the project—with the "eye" design being the first candidate. But considering this a bit too surreal, the next candidate designs were more staid. The title still wasn't settled, but in the fall of 1992 a few covers were tried:

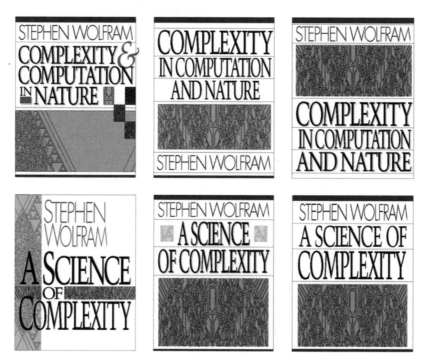

I thought these covers looked a bit drab, so we brightened them up, and by 1993—and after a few "color explorations"

we had a "working cover" for the book (complete with its working title), carrying over typography from the previous designs, but now featuring an image of rule 30 together with the "mascot of the project": a textile cone shell with a rule-30-like pigmentation pattern:

When I changed the title in 1994, the change was swiftly executed on the cover—with my draft copy from the time being a charming palimpsest with *A New Kind of Science* pasted over *A Science of Complexity*:

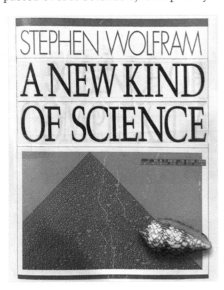

I was never particularly happy with this cover, though. I thought it was a bit "static", particularly with all those boxed-in elements. And compared to other "popular books" in bookstores at the time, it was a very "quiet" cover. My book designer tried to "amp it up"

sometimes still with a hint of mollusc:

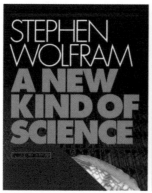  

"Not that loud!", I said. So he quietened it down, but now with the type getting a bit more dynamic:

Then a bit of a breakthrough: just type and cellular automaton (now rule 110):

It was nice and simple. But now it seemed perhaps too quiet. We punched up the type, just leaving the cellular automaton as a kind of decoration:

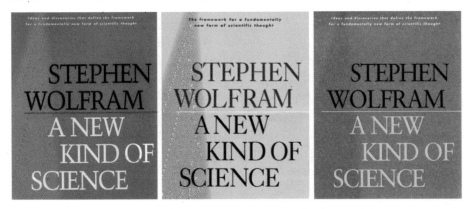

And there were a variety of ways to handle the type (maybe even with an emphasized subtitle—complete with a designer's misspelling):

But the important point was that we'd basically backed into an idea: why not just use the natural angles of the structures in rule 110 to delimit the cellular automaton on the cover? As so often happens, the computational universe had "spontaneously" thrown up a good idea that we hadn't thought of.

I didn't think the cover was quite "there", but it was making progress. Right around this time, though, we were in discussions with a big New York publisher about them publishing the book, and they were trying to sell us on the value they could add. They were particularly keen to show us their prowess at cover design. We patiently explained that we had quite a large and good art department, which happened to have even recently won some national awards for design.

But the publisher was sure they could do better. I remember saying: "Go ahead and try"— and then adding, "But please don't show us something from someone who has no idea what kind of book this is."

Several weeks later, with some fanfare, they produced their proposal:

Yup, mollusc shells can be found on beaches. But this wasn't a "beach-reading novel" kind of book. And it would be an understatement to say we weren't impressed.

So, OK, it was on us: as I'd expected, we'd have to come up with a cover design. My notes aren't dated, but sometime around then I started thinking harder about the design myself. I was playing around with rule 30, imagining a "physicalized" version of it (with 3D, letters casting shadows, etc.):

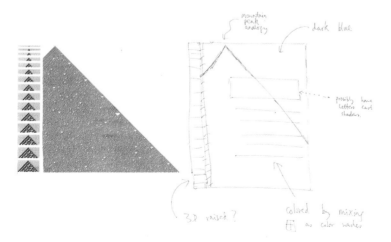

I find in my archives some undated sketches of further "physicalized" cover concepts (or, at least I assume they were cover concepts, and, yes, sadly I've never learned to draw, and I can't even imagine who that dude was supposed to be):

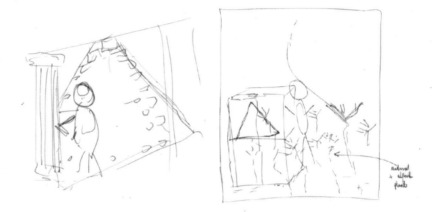

But then we had an idea: maybe the strangely shaped triangle could be like a shaft of light illuminating a cellular automaton image. We talked about the metaphor of the science "providing illumination". I was very taken with the notion that the basic ideas of the science could have been discovered even in ancient times. And that made us think about cellular automaton markings in a cave, suddenly being illuminated by an archaeologist's flashlight. But how would we make a picture of something like that?

We tried some "stone effects":

We investigated finding a stone mason who could carve a cellular automaton pattern into something like a gravestone. (3D printing wasn't a thing yet.) We even tried some photographic experiments. But with the cellular automaton pattern itself having all sorts of fine detail, one barely even noticed a stone texture. And so we went back to pure computer graphics, but now with a "shaft of light" motif:

It wasn't quite right, but it was getting closer. Meanwhile, the New York publisher wanted to have another try. Their new, "spiffier" proposal (offering type alternatives for "extra credit") was:

(The shell, now shrunk, was being kept because their sales team was enamored of the idea of a tie-in whereby they would give physical shells to bookseller sales prospects.)

OK, so how were we going to tune up the cover? The cellular automaton triangle wasn't yet really looking much like a shaft of light. It was something to do with the edges, we thought:

It was definitely very subtle. We tried different angles and colors:

We tried, and rejected, sans serif, and even partial sans serif:

And by July 1995 the transition was basically complete, and for the first time our draft printouts started looking (at least on the outside) very much like modern NKS books:

Specifying just what color should be printed was pretty subtle, and over the months that followed we continued to tweak, particularly the "shaft of light"

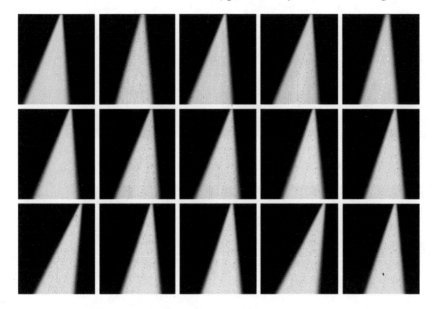

until eventually *A New Kind of Science* got its final cover:

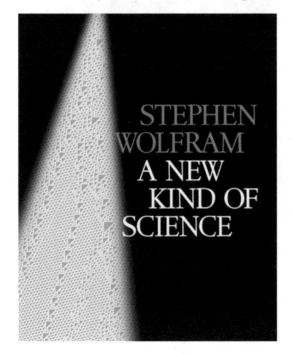

The Making of *A New Kind of Science*

All along we'd also been thinking about what would show up on the spine of the book—and occasionally testing it in an "identity parade" on a bookshelf. And as soon as we had the "shaft of light" idea, we immediately thought of it wrapping around onto the spine:

Part of what makes the cover work is the specific cellular automaton pattern it uses—which, in characteristic form, I explained in the notes (and, yes, the necessary initial conditions were found by a search, and are now in the Wolfram Data Repository):

## The Opening Paragraphs

How should the NKS book begin? When I write something I always like to start writing at the beginning, and I always like to say "up front" what the main point is. But over the decade that I worked on the NKS book, the "main point" expanded—and I ended up coming back and rewriting the beginning of the book quite a few times.

In the early years, it was pretty much all about complexity—though even in 1991 the term "a new kind of science" already makes an appearance in the text:

1991

### Goals

Complexity is a common phenomenon. Every day, in fact, we see many systems whose form or behavior shows complexity that seems to defy simple description.

What does science have to say about such systems?

There are certainly many scientific results about the components that make up such systems. But traditional science has had surprisingly little to say about how these components act together to produce the complexity we see.

My goal in this book is to introduce a new kind of science that specifically concentrates on the phenomenon of complexity.

It is not at first clear that it is possible to develop a coherent science of this kind. After all, the details of different systems that show complexity are often quite different.

1991

### The Science of Complexity

Complexity is a common phenomenon. Every day, in fact, we see many systems whose form or behavior shows complexity that seems to defy simple description.

What does science have to say about such systems? There are certainly many scientific results about the components that make up such systems. But traditional science has had surprisingly little to say about how these components act together to produce the complexity we see.

My goal in this book is to introduce a new kind of science that specifically concentrates on the phenomenon of complexity. It is not at first clear that it is possible to develop a coherent science of this kind. After all, the details of different systems that show complexity are often quite different.

1992

### The Science of Complexity

Complexity is very common. Every day in fact we see many systems whose form or behavior is so complex that it seems to defy any simple description.

Quite often much is known about the individual components that make up these systems. But typically rather little is known about how these components act together to produce the complexity we see. And although the details of the components in different systems are usually quite different, the phenomenon of complexity is surprisingly universal.

The main purpose of this book is to find the fundamental mechanisms by which such complexity is produced. The results apply to a very wide range of systems, physical, biological, social and otherwise, and in many cases they shed new light on fundamental

1992

In almost every area of science, technology and business, complexity is an important issue. Whether one is concerned with turbulence in the early universe, with the growth of biological organisms, or with the fluctuations of prices in the market, one is continually confronted with behavior of great complexity.

One might have thought that science would long ago have developed a basic theory for such a widespread phenomenon. But in fact science has in the past had remarkably little to say about complexity. Indeed, basic science has traditionally tended to concentrate on just those situations where direct confrontation with complexity can be avoided.

In 1993, I considered a more "show, don't tell" approach that would be based on photographs of simple and complex forms:

1993

### The Phenomenon of Complexity

When we look at the natural world, we cannot fail to be struck by the complexity of what we see. But why is it that there is so much complexity in nature?

It turns out that despite all their successes over the past few centuries the existing sciences have never had much to say about this very fundamental question. My goal in this book, however, is to develop a

1994

### Complexity in Nature

My goal in this book is not a modest one. I want to find a way to understand some very basic aspects of the natural world that science has never before been capable of handling.

When we look at the natural world, we cannot fail to be struck by the complexity of what we see. But what is the fundamental origin of this complexity?

Despite their great successes over the past few centuries the existing sciences have never had much to say about this very central ques-

But soon the pictures were gone, and I began to concentrate more on how what I was doing fitted into the historical arc of the development of science—though still under a banner of complexity:

1995

**Complexity in Nature**

Over the past several centuries science has had many spectacular successes. Yet despite these successes there are still many fundamental questions about the natural world that remain entirely unanswered.

At a very basic level, for example, we do not even know why so many of the systems that we see in nature appear to us complex. After all, it could be that nature would mostly generate only shapes such as squares and circles that we consider simple. But in practice we find that nature instead produces a vast range of much more complex forms.

Our own experience in building things tends to make us assume

After my 1996 hiatus (spent finishing Mathematica 3.0) the text of the opening section hadn't changed, but the title was now "The Need for a New Kind of Science":

1997

**The Challenge|Mystery of Complexity** [1.1]

Over the past several centuries science has had many spectacular successes. Yet despite these successes there are still many fundamental questions about the natural world that remain entirely unanswered.

At a very basic level, for example, we do not even know why so many of the systems that we see in nature appear to us complex. After all, it could be that nature would mostly generate only shapes such as squares and circles that we consider simple. But in practice we find that nature instead produces a vast range of much more complex forms.

And I was soon moving further away from complexity, treating it more as "just an important example":

1998

**The Challenge|Mystery of Complexity|What Science Has Left Undone|Going Beyond Existing Science** [1.1]

Over the past several centuries science has had many spectacular successes. Yet despite these successes there are still many fundamental questions about the natural world that remain entirely unanswered.

[A remarkably large fraction of these revolve around the very basic question of why so many of the systems that we see in nature appear to us so complex. It could be, after all, xxx]

At a very basic level, for example, we do not even know why so many of the systems that we see in nature appear to us so complex.

Then, in 1999, "complexity" drops out of the opening paragraphs entirely, and it becomes all about methodology and the arc of history:

1999

### The Foundations of a Revolution [1.1]

Three centuries ago science was transformed by the dramatic new idea that rules based on mathematical equations could be used to describe the natural world. My purpose in this book is to initiate another such transformation, and to introduce a new kind of science that is based on the much more general types of rules that can be embodied in simple computer programs.

It has taken me the better part of twenty years to build the intellectual structure that is needed, but I have been amazed by its results. For what I have found is that with the new kind of science I

And in fact from there on out the first couple of paragraphs don't change—though the section title softens, taking out the explicit mention of "revolution":

2001

### An Outline of Basic Ideas

Three centuries ago science was transformed by the dramatic new idea that rules based on mathematical equations could be used to describe the natural world. My purpose in this book is to initiate another such transformation, and to introduce a new kind of science that is based on the much more general types of rules that can be embodied in simple computer programs.

It has taken me the better part of twenty years to build the intellectual structure that is needed, but I have been amazed by its results. For what I have found is that with the new kind of science I

It's interesting to notice that even though until perhaps 1998 before the opening of the book reflected "moving away from complexity", other things I was writing already had. Here, for example, is a candidate "cover blurb" that I wrote on January 11, 1992 (yes, a decade early):

> **Computation and the Complexity of Nature**
>
> Computation has already had a great impact on almost every area of science. But this book argues that the still greater impact of computation on science is yet to come. Beyond using computers as practical tools, one can use computation as the basis for a whole new conceptual framework for studying some of the most fundamental problems of science.
>
> Central to this framework is the idea that natural processes should be viewed as computations. This book shows that nature need use only some of the simplest imaginable computer programs to produce the kind of complexity we see in many physical, biological and other systems, but that science has traditionally been unable to explain.
>
> The book offers solutions to some long-standing problems in science, such as the validity of Second Law of thermodynamics, the origin of fluid turbulence and the complexity of biological forms. It also challenges ideas such as chaos as the source of randomness in nature, the independence of the observer in classical physics, as well as the role of traditional mathematics in the description of natural processes.
>
> This is a book about fundamental ideas, and it is written to communicate these ideas as clearly as possible, without the technical jargon of any particular field of science. Illustrated on almost every page by extensive computer graphics, the ideas presented in this book should be important to all those interested in science or scientific thought.

And as I pull this out of my archives, I notice at the bottom of it:

> Also in progress: A Guide to Complexity in Cellular Automata and Other Simple Systems (400 pp. ?)

Hmm. That would have been interesting. But another 400 pages?

## Ten Years of Writing

By the end of 1991 the basic concept of what would become *A New Kind of Science* was fairly clear. At the time, I still thought—as I had in the 1980s—that the best "hook" was the objective of "explaining complexity". But I perfectly well understood that from an intellectual and methodological point of view the most important part of the story was that I was starting to truly take seriously the notion of computation—and starting to think broadly in a fundamentally computational way.

But what could be figured out like this? What about systems based on constraints? What about systems that adapt or learn? What about biological evolution? What about fundamental physics? What about the foundations of mathematics? At the outset, I really didn't know whether my approach would have anything to say about these things. But I thought I should at least try to check each of them out. And what happened was that every time I turned over a (metaphorical) rock it seemed like I discovered a whole new world underneath.

It was intellectually exciting—and almost addictive. I would get into some new area and think "OK, let me see what I can figure out here, then move on". But then I would get deeper and deeper into it, and weeks would turn into months, and months would turn into years. At the beginning I would sometimes tell people what I was up to. And they would say "That sounds interesting. But what about X, Y, Z?" And I would think "I might as well try and answer those questions too". But I soon realized that I shouldn't be letting myself get distracted: I already had more than enough very central questions to answer.

And so I decided to pretty much "go hermit" until the book was done. An email I sent on October 1, 1992, summarizes how I was thinking at the time:

> **Date:** Thu, 1 Oct 1992 12:23:30
> **From:** swolf@wri.com
>
> I'm afraid I'm still working hard to ignore the world.....
> I'm currently doing a rather ambitious science project that I hope will be finished by the middle of next year. After that, more technology.

But that email was right before I discovered yet more kinds of computational systems to explore, and before I'd understood applications to biology, and physics, and mathematics, and so on.

In the early years of the project I'd had various "I could do that as well" ideas. In 1991 I thought about dashing off an *Introduction to Computing* book (maybe I should do that now!). In 1992 I had a plan for creating an email directory for the world (a very proto LinkedIn). In 1993 I thought about TIX: "The Information Exchange" (a proto web for computable documents).

But thinking even a little about these things basically just showed me how much what I really wanted to do was move forward on the science and the book. I was still energetically remote-CEOing my company. But every day, by mid-evening, I would get down to science, and work on it through much of the night. And pretty much that's how I spent the better part of a decade.

My personal analytics data of outgoing emails show that during the time I was working on the book I became increasingly nocturnal (I shifted and "stabilized" after the book was finished):

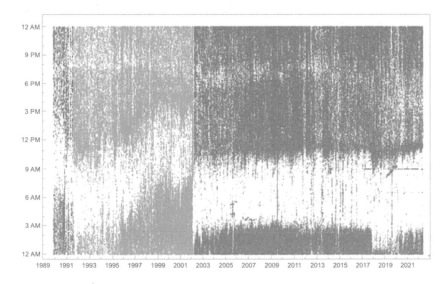

I had started the NKS book right after the big push to release Mathematica 2.0. And thinking the book would take a year or maybe 18 months I figured it would be long finished before there was a new version of Mathematica, and another big push was needed. But it was not to be. And while I held off as long as I could, by 1996 there was no choice: I had to jump into finishing Mathematica 3.0.

From the beginning until now I've always been the ultimate architect of what's now the Wolfram Language. And back in the 1990s my way of defining the specification for the language was to write its documentation, as a book. So getting Mathematica 3.0 out required me writing a new edition of *The Mathematica Book*. And since we were adding a lot in Version 3, the book was long—eventually clocking in at 1403 pages. And it took me a good part of 1996 to write it.

But in September 1996, Mathematica 3.0 was released, and I was able to go back to my intense focus on science and the NKS book. In many ways it was exhilarating. With Wolfram Language as a tool, I was powering through so much research. But it was difficult stuff. And getting everything right—and as clear as possible—was painstaking, if ultimately deeply satisfying, work. On a good day I might manage to write one page of the book. Other times I might spend many days working out what would end up as just a single paragraph in the notes at the back of the book.

I kept on thinking "OK, in just a few months it'll be finished". But I just kept on discovering more and more. And finding out again and again that sections in the table of contents that I thought would just be "quick notes" actually led to major research projects with all sorts of important and unexpected results.

A 1995 picture captured my typical working setup:

A year or so later, I had the desk I'm still sitting at today (though not in the same location), and a (rarely used) webcam had appeared:

A few years after that, the computer monitor was thinner, two young helpers had arrived, and I was looking distinctly unkempt and hermit-like:

In 2000 a photographer for Forbes captured my "caged scientist" look

along with a rather nice artistically lit "still life" of my working environment (complete with a "from-the-future" thicker-than-real-life mockup of the NKS book):

But gradually, inexorably, the book got closer and closer to being finished. The floor of my office had been covered with piles of paper, each marked with whatever issue or unfinished section they related to. But by 2001 the piles were disappearing—and by the fall of that year they were all but gone: a visible sign that the book was nearing completion.

### Tracking Everything Down: A Decade of Scholarship

*A New Kind of Science* is—as its title suggests—a book about new things. But an important part of explaining new things is to provide context for them. And for me a key part of the context for things is always the story of what led to them. And that was something I wanted to capture in the NKS book.

Typically there were two parts: a personal narrative of how I was led to something—and a historical narrative of what in the past might connect to it. The academic writing style that I'd adopted in the 1980s really didn't capture either of these. So for the NKS book I needed a new style. And there were again two parts to this. First, I needed to "put myself into the text", describing in the first person how I'd reached conclusions, and what their importance to me was. And second, I needed to "tell the story" of whatever historical developments were relevant.

Early on, I made the decision not to mix these kinds of narratives. I would talk about my own relation to the material. And I would talk about other people and their historical relation to the material. But I didn't talk about my interactions with other people. And, yes, there are lots of wonderful stories to tell—which perhaps one day I'll have a chance to systematically write down. But for the NKS book I decided that these stories—while potentially fun to read—just weren't relevant to the absorption and contextualization of what I had to say. So, with a bit of regret, I left them out.

In typical academic papers one references other work by inserting pure, uncommented citations to it. And deep within some well-developed field, this is potentially an adequate thing to do. Because in such a field, the structure is in a sense already laid out, so a pure citation is enough to explain the connection. But for the NKS book it was quite different. Because most of the time the historical antecedents were necessarily done in quite different conceptual frameworks—and typically the only reasonable way to see the connection to them was to tell the story of what was done and why, recontextualized in an "NKS way".

And what this meant was that in writing the NKS book, I ended up doing a huge amount of "scholarship", tracking down history, and trying to piece together the stories of what happened and why. Sometimes I personally knew—or had known—the people involved. Sometimes I was dealing with things that had happened centuries ago. Often there were mysteries involved. How did this person come to be thinking about this? Why didn't they figure this-or-that out? What really was their conceptual framework?

I've always been a person who tries to "do my homework" in any field I'm studying. I want to know both what's known, and what's not known. I want to get a sense of the patterns of thinking in the field, and "value systems" of the field. Many times in working on the NKS book I got the sense that this-or-that field should be relevant. But what was important for the NKS book was often something that was a footnote—or was even implicitly ignored—by the field. And it also didn't help that the names for things in particular fields were often informed by their specific uses there, and didn't connect with what was natural for the NKS book.

I started the NKS book shortly after the web was invented, and well before there was substantial content on it. So at least at first a lot of my research had to be done the same way I'd done it in the 1980s: from printed books and papers, and by using online and printed abstracting systems. Here's part of a "search" from 1991 for papers with the keyword "automata":

By the end of writing the NKS book I'd accumulated nearly 5000 books, a few of them pictured here in their then-habitat circa 1999 (complete with me at my I've-been-on-this-project-too-long lifetime-maximum weight):

I had an online catalog of all my books, which I put online soon after the NKS book was published. I also had file cabinets filled with more than 7000 papers. Perhaps it might have been nice when the NKS book was published to be able to say in a kind of traditional academic style "here are the 'citations'" (and, finally, 20 years later we're about to be able to actually do that). But at the time it wasn't the simple citations I wanted, or thought would be useful; it was the narrative I could piece together from them.

And sometimes the papers weren't enough, and I had to make requests from document archives, or actually interview people. It was hard work, with a steady stream of surprises. For example, in Stan Ulam's archives we found a (somewhat scurrilous) behind-the-scenes interaction about me. And after many hours of discussion John Conway admitted to me that his usual story about the origin of the Game of Life wasn't correct—though I at least found the true story much more interesting (even if some mystery still remains). There were times when the things I wanted to know were still entangled in government or other secrecy. And there were times when people had just outright forgotten, often because the things I now cared about just hadn't seemed important before—and now could only be recovered by painstakingly "triangulating" from other recollections and documents.

There were so many corners to the scholarship involved in creating the NKS book. One memorable example was what we called the "People Dates" project. I wanted the index to include not only the name of every person I mentioned in the book, but also their dates, and the primary country or countries in which they worked, as in "Wolfram, Stephen (England/USA, 1959– )."

For some people that information was straightforward enough to find. But for other people there were challenges. There were 484 people altogether in the index, with a roughly exponentially increasing number born after about 1800:

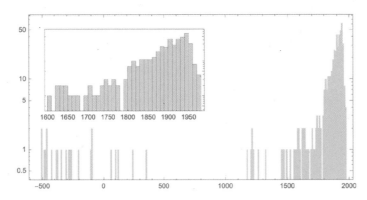

For ones who were alive we just sent them email, usually getting helpful (if sometimes witty) responses. In other cases we had to search government records, ask institutions, or find relatives or other personal contacts. There were lots of weird issues about transliterations, historical country designations, and definitions of "worked in". But in the end we basically got everything (though for example Moses Schönfinkel's date of death remained a mystery, as it does even now, after all my recent research).

Most of the historical research I did for the NKS book wound up in notes at the back of the book. But of all the 1350 notes spread over 348 small-print pages, only 102 were in the end historical. The other notes covered a remarkable range of subject matter. They provided background information, technical details and additional results. And in many ways the notes represent the highest density of information in the NKS book—and I, for example, constantly find myself referring to them, and to their pithy (and, I think, rather clear) summaries of all sorts of things.

When I was working on the book there were often things I thought I'd better figure out, just in case they were relevant to the core narrative of the book. Sometimes they'd be difficult things, and they'd take me—and my computers—days or even weeks. But quite often what came out just didn't fit into the core narrative of the book, or its main text. And so the results were relegated to notes. Maybe there'll just be one sentence in the notes making some statement. But behind that statement was a lot of work.

Many times I would have liked to have had "notes to the notes". But I restrained myself from adding yet more to the project. Even though today I've sometimes found myself writing even hundreds of pages to expand on what in the NKS book is just a note, or even a part of a note.

The 1990s spanned the time from the very beginning of the web to the point where the web had a few million pages of content. And by the later years of the project I was making use of the web whenever I could. But often the background facts I needed for the notes were so obscure that there was nothing coherent about them on the web—and in fact even today it's common for the notes to the NKS book to be the best summaries to be found anywhere.

I figured, though, that the existence of the web could at least "get me off the hook" on some work I might otherwise have had to do. For example, I didn't think there was any point in giving explicit citations to documents. I made sure to include relevant names of people and topics. Then it seemed as if it'd be much better just to search for those on the web, and find all relevant documents, than for me to do all sorts of additional scholarship trying to pick out particular citations that then someone might have to go to a library to look up.

## Finishing the Book

I'm not sure when I could say that the finishing of the NKS book finally seemed in sight. We'd been making bound book mockups since early 1994. Looking through them now it's interesting to see how different parts gradually came together. In July 1995, for example, there was already a section in Chapter 9 on "The Nature of Space", but it was followed by a section on the "Nature of Time" that was just a few rough notes. There's a hiatus in mockups in 1996 (when I was working on Mathematica 3.0) but when the mockups pick up again in January 1997—now bound in three volumes—there's a section on "The Nature of Time" containing an early (and probably not very good) idea based on multiway systems that I'd long since forgotten (later "The Nature of Time" section would be broken into different sections):

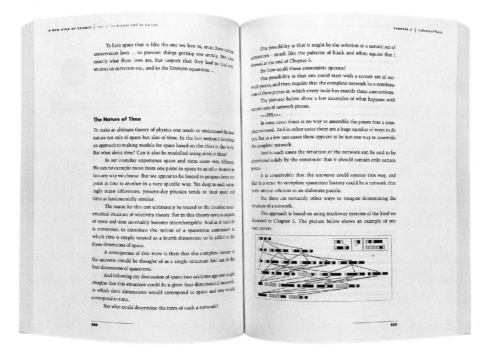

Already in 1997 there's a very rough skeleton of Chapter 12—with a fairly accurate collection of section headings, but just 18 pages of rather rough notes as content. Meanwhile, there's a post-Chapter-12 "Epilog" that sprouts up, to be dropped only late in the project (see below). Chapter 12 begins to "bulk up" in late 1999, and in 2000 really "takes off", for example adding the long section on "Implications for the Foundations of Mathematics". At that point our rate of making book mockups began to pick up. We'd been indicating different mockups with dates and colored labeling ("the banana version", etc.) But, finally, dated February 14, 2001, there's a version labeled (in imitation of software release nomenclature) "Alpha 1".

And by then I was starting to make serious use of the machinery for doing large projects that we'd developed for so many years at Wolfram Research. The "NKS Project" started having project managers, build systems and internal websites (yes, with garish web colors of the time):

**Preface and Chapter 1**

| Title | Starting Page Number | Pages stable | Text DQA checked | DQA changes implemented | Indexed | Graphics regenerated | Microtweaked | Layout/graphics reviewed by andre | Graphics DQA checked | Layout/graphics reviewed by sw | Comments |
|---|---|---|---|---|---|---|---|---|---|---|---|
| Preface | ix | 05-01-01 | | | completed 11-22-01 | no graphics | | | | | |
| Chapter 1-The Foundations for a New Kind of Science | 1 | 05-01-01 | 06-05-01; 01-30-02 | 06-08-01; 01-30-02 | completed 11-22-01 | no graphics | | completed 09-01-01 | completed 12-18-01 | completed 10-04-01 | completed 12-18-01 | |
| 1.1 An Outline of Basic Ideas | 1 | 05-01-01 | 06-05-01; 01-30-02 | 06-08-01; 01-30-02 | completed 11-22-01 | no graphics | | completed 09-01-01 | completed 12-18-01 | completed 10-04-01 | completed 12-18-01 | |
| 1.2 Relations to Other Areas | 7 | 05-01-01 | 06-05-01; 01-30-02 | 06-08-01; 01-30-02 | completed 11-22-01 | no graphics | | completed 09-01-01 | completed 12-18-01 | completed 10-04-01 | completed 12-18-01 | |
| 1.3 Some Past Initiatives | 12 | 05-01-01 | 06-05-01; 01-30-02 | 06-08-01; 01-30-02 | completed 11-22-01 | no graphics | | completed 09-01-01 | completed 12-18-01 | completed 10-04-01 | completed 12-18-01 | |
| 1.4 The Personal Story of the Science in This Book | 17 | 05-01-01 | 06-05-01; 01-30-02 | 06-08-01; 01-30-02 | completed 11-22-01 | no graphics | | completed 09-01-01 | completed 12-18-01 | completed 10-04-01 | completed 12-18-01 | |

**Chapter 2**

| Title | Starting Page Number | Pages stable | Text DQA checked | DQA changes implemented | Indexed | Graphics regenerated | Microtweaked | Layout/graphics reviewed by andre | Graphics DQA checked | Layout/graphics reviewed by sw | Comments |
|---|---|---|---|---|---|---|---|---|---|---|---|
| Chapter 2-The Crucial Experiment | 23 | 03-28-01 | 05-21-01; 08-14-01; 01-09-02 | 05-30-01; 08-29-01; 01-12-02 | completed 11-22-01 | completed prior to 05-04-01 jayw | completed 09-01-01 | completed 12-18-01 | completed 10-04-01 | completed 12-18-01 | |
| 2.1 How Do Simple Programs Behave? | 23 | 03-28-01 | 05-21-01; 08-14-01; 01-09-02 | 05-30-01; 08-29-01; 01-12-02 | completed 11-22-01 | completed prior to 05-04-01 jayw | completed 09-01-01 | completed 12-18-01 | completed 10-04-01 | completed 12-18-01 | |
| 2.2 The Need for a New Intuition | 39 | 03-28-01 | 05-21-01; 08-14-01; 01-09-02 | 05-30-01; 08-29-01; 01-12-02 | completed 11-22-01 | completed prior to 05-04-01 jayw | completed 09-01-01 | completed 12-18-01 | completed 10-04-01 | completed 12-18-01 | |
| 2.3 Why These Discoveries Were Not Made Before | 42 | 03-28-01 | 05-21-01; 08-14-01; 01-09-02 | 05-30-01; 08-29-01; 01-12-02 | completed 11-22-01 | completed prior to 05-04-01 jayw | completed 09-01-01 | completed 12-18-01 | completed 10-04-01 | completed 12-18-01 | |

We'd had the source for the book in a source control system for several years, but as far as I was concerned the ultimate source for the book was my filesystem, and a specific set of directories that, yes, are still there in my filesystem all these years later:

| Name | Date Modified | Size | Kind |
| --- | --- | --- | --- |
| ▶ Archives | Dec 26, 1999 at 7:33 AM | -- | Folder |
| ▶ Checking | Mar 12, 2001 at 12:00 AM | -- | Folder |
| ▶ Converted | Mar 13, 2001 at 3:12 PM | -- | Folder |
| ▶ Dictionaries | Jan 29, 2002 at 4:12 PM | -- | Folder |
| ▶ Extras | May 14, 2000 at 2:29 AM | -- | Folder |
| ▶ Graphics | Apr 30, 2001 at 1:16 PM | -- | Folder |
| ▶ GraphicsArchive | Feb 21, 2001 at 2:51 PM | -- | Folder |
| ▶ Images | May 11, 2001 at 1:07 PM | -- | Folder |
| ▶ NewText | Sep 14, 1998 at 5:58 PM | -- | Folder |
| ▶ Notebooks | Jun 12, 2001 at 10:48 PM | -- | Folder |
| ▶ NotesCode | Dec 5, 2019 at 11:51 PM | -- | Folder |
| ▶ Outtakes | Dec 27, 2003 at 11:56 PM | -- | Folder |
| ▶ PDFImages | Apr 13, 2006 at 12:33 PM | -- | Folder |
| ▶ PictorialSummary | Nov 12, 1999 at 5:40 AM | -- | Folder |
| ▶ Results | Jun 12, 2001 at 10:50 PM | -- | Folder |
| ▶ Scratch | Feb 9, 2001 at 3:12 PM | -- | Folder |
| ▶ Text | Feb 28, 2002 at 12:50 PM | -- | Folder |

Everything was laid out by chapter and section. **Text** contained the FrameMaker files. **Notebooks** contained the source notebooks for all the diagrams (with long-to-compute results pre-stored in **Results**):

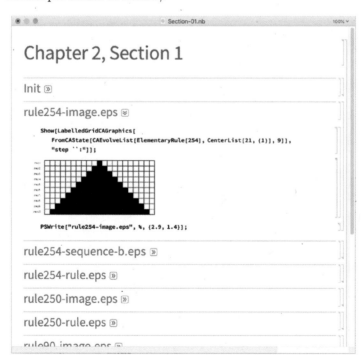

The Making of *A New Kind of Science*

The workflow was that every diagram was created in Wolfram Language, then saved as an EPS file. (EPS or "Encapsulated PostScript" was a forerunner of PDF.) And gradually, over the course of years, more and more EPS files were generated, here reconstructed in the order of their generation, starting around 1994:

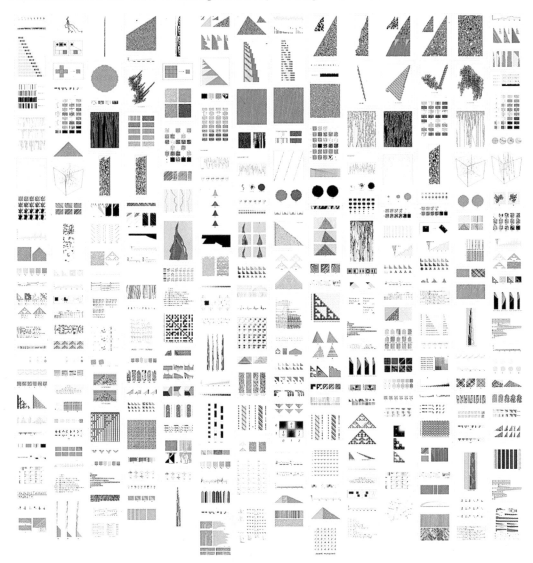

In creating all these EPS files, there was lots of detailed tweaking done, for example in the exact (programmatically specified) sizes for the images given in the files. We'd built up a whole diagram-generating system, with all sorts of detailed standards for sizings and spacings and so on. And several times—particularly as a result of discovering quirks in the printing process—we decided we had to change the standards we were using. This could have been a project-derailing disaster. But because we had everything programmatically set up in notebooks it was actually quite straightforward to just go through and automatically regenerate the thousand or so images in the book.

Each EPS file that was generated was put in a Graphics directory, then imported ("by reference") by FrameMaker into the appropriate page of the book. And the result was something that looked almost like the final NKS book. But there were two "little" wrinkles, that ended up leading to quite a bit of technical complexity.

The first had to do with the fragments of Wolfram Language code in the notes. At the time it was typical to show code in a simple monospaced font like Courier. But I thought this looked ugly—and threw away much of the effort I'd put into making the code as elegant and readable as possible. So I decided we needed a different code font, and in particular a proportionally spaced sans serif one. But there was a technical problem with this. Many of the characters we needed for the code were available in any reasonable font. But some characters were special to the Wolfram Language—or at least were characters that for example we'd been responsible for being included in the Unicode standard, and weren't yet widely supported in fonts.

And the result was that in addition to all the other complexities of producing the book we had to design our own font, just for the book:

**NewScienceSans-Italic**

But that wasn't all. In Mathematica 3.0 we had invented an elaborate typesetting system which carefully formatted Wolfram Language code, breaking it into multiple lines if necessary. But how were we to weave that nicely formatted code into the layouts of pages in FrameMaker? In the end we had to use Wolfram Language to do this. The way this worked is that first we exported the whole book from FrameMaker in "Maker Interchange Format" (MIF). Then we parsed the resulting MIF file in Wolfram Language, in effect turning the whole book into a big symbolic expression. At that point we could use whatever Wolfram Language functionality we wanted, doing various pattern-matching-based transformations and typesetting each of the pieces of code. (We also handled various aspects of the index at this stage.) Then we took the symbolic expression, converted it to MIF, and imported it back into FrameMaker.

In the end the production of the book was handled by an automated build script—just like the ones we used to build Mathematica (the full build log is 11 pages long):

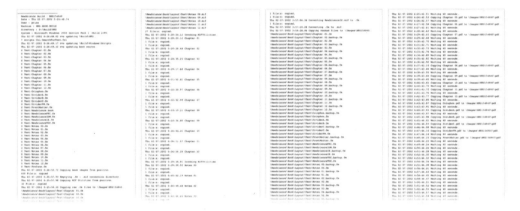

But, OK, so by early 2001 we were well on the way to setting all these technical systems up. But there was more to do in "producing the book"—as indicated for example by the various column headings in the project management internal website. "Graphics regenerated" was about regenerating all the EPS files with the final standards for the book. "Microtweaking" was about making sure the placement of all the graphics was just right. Then there were various kinds of what in our company we call "document quality assurance", or DQA—checking every detail of the document, from grammar and spelling to overall consistency and formatting. (And, yes, developing a style guide that worked with my sometimes-nonstandard—but I believe highly sensible!—writing conventions.)

In addition to checking the form of the book, there was also the question of checking the content. Much of that—including extensive fact checking, etc.—had gone on throughout the development of the book. But near the end one more piece of checking had to do with the code that was included in the book itself. Our company has had a long history of sophisticated software quality assurance ("SQA"), and I applied that to the book—for example having extensive tests written for all the code in the book.

Much like for software, once we reached the first "Alpha version" of the book we also started sending it out to external "alpha testers"—and got a modest but helpful collection of responses. We had several pages of instructions for our "testers" (that we called "readers" since, after all, this was a book):

---

### Things we'd like from readers of *A New Kind of Science* at this time:

**Blatant Errors**

The main thing we want to do is to find out about things that are definitely wrong. Examples include:
- typos (spelling errors, words missed out, etc.)
- "*facts*" that are categorically wrong (see below)
- inconsistencies (different terminology for the same thing)
- glitches in pictures (little uglinesses or inconsistencies; they don't contain what they are described as containing, etc.)
- typography problems (wrong or ugly fonts, things crashing into each other, etc.)
- infrastructural problems (incorrect page references, notes out of order, etc.)
- index problems (see below)

Please mention **anything** that seems definitely wrong. Do not assume that someone else will notice whatever it is you've noticed. Do not assume that things that look very wrong must be right, because someone else would surely have noticed it if it was that wrong. They may think the same as you, and in the end we may miss something big.

Having said that, please note that the focus is on finding things that are **definitely wrong**. If things are a "*matter of opinion*", then probably Stephen Wolfram thought about them when he wrote what he wrote, and in the interests of getting this huge project finished, we'll almost certainly just want to leave them the way they are.

**Fact-checking**

Particularly in the notes, there are a huge number of "*conventional wisdom*" facts quoted. We want to check as many of these as we can. If you know references for particular facts, tell them to us. If you think a fact is wrong, please try to check carefully whether this is really the case. If you try to check, but don't succeed, tell us about it, and with luck someone else can do the checking.

We're particularly concerned about notes that common experience or basic historical knowledge show are wrong. Examples (not actually in the book!) would be: "bananas start off as spherical fruits" or "as Isaac Newton showed in 1734" (he was dead by then).

Also worth pointing out are elements of history that are likely to offend particular groups or individuals. (BC isn't changing to BCE, and political correctness won't be overdone.)

You don't need to comment on anything that looks like it would be filled in when an XXXX is turned into something specific. If the XXXX refers to some kind of fact, then by all means tell us if you know what the fact is.

The book quite often says "see page ____", particularly in the notes. Think about whether these references are correct, and whether there might perhaps be a few additional ones.

After the "Alpha 1" version of the book in February 2001, there followed six more "Alpha" versions. In "Alpha 1" there were still XXXX's scattered around the text, alignment and other issues in graphics—and some of the more "philosophical" sections in the book were just in note form, crossed out with big X's in the printout. But in the course of 2001 all these issues got ironed out. And on January 15, 2002, I finished and dated the preface.

Then on February 4, 2002, we produced the "Beta 1" version of the book—and began to make final preparations for its printing and publication. It had been a long road, illustrated by the sequence of intermediate versions we'd generated, but we were nearing the end:

## The Joy of Indexing

I like indices, and the index to the NKS book—with its 14,967 entries—is my all-time favorite. In these times of ubiquitous full-text search one might think that a book index would just be a quaint relic of the past (and indeed some younger people don't even seem to know that most books have indices!). But it definitely isn't with the NKS book. And indeed when I want to find something in the book, the place I always turn first is the index (now online).

I started creating the index to the NKS book in the spring of 1999, and finished it right before the final version of the book was produced in February 2002. I had already had the experience of creating indices to five editions of *The Mathematica Book*, and had seen the importance of those indices in people's actual use of Mathematica. I had developed various theories about how to make a good index—which sometimes differed from conventional wisdom—but seemed to work rather well.

A good index, I believe, should list whatever terms one might actually think of looking up, regardless of whether it's those literal terms—or just synonyms for them—that appear in the text. If there's a phrase (like "finite automata") explicitly list it in all the ways people might think of it ("finite automata", "automata, finite"), rather than having some "theory" (that the users of the index are very unlikely to know) about how to list the phrase. And perhaps most important, generously include subterms, "subdividing" until each individual entry references at most a few pages. Because when you're looking for something, you want to be able to zero in on a particular page, not be confronted with lots of "potentially relevant" pages. And well-chosen subterms immediately give a kind of pointillistic map of the coverage of some area.

I've always enjoyed creating indices. For me it's an interesting exercise in quickly organizing knowledge and identifying what's important, as well as engaging in rapid "what are different ways to say that?" association. (And, yes, a similar skill is needed in linguistic curation for the natural language understanding system of Wolfram|Alpha.) For the NKS book (and other indices) my basic strategy was to go through the book page by page, adding tags for index entries. But what about consistency? Did I just index "Fig leaves" in one place, and somewhere else index "Leaves, fig" instead? We built Wolfram Language code to identify such issues. But eventually I just generated the alphabetical index, and read through it. And then had Wolfram Language code that could realign tags to correct the source of whatever fixes I made—which most often related to subterms.

At first I broke the index into an ordinary "Index" and an "Index of Names". But what counted as a "name"? Only a person's name? Or also a place name? Or also "rule 30"? Within a couple of months I had combined everything into an "Index of words, names, concepts and systems"—which soon became headed just "Index" (with a pointer to a note about what was in it).

The final index is remarkably eclectic—reflecting of course the content of the book. After "Field theory (physics)" comes "Fields (agricultural)", followed by "Fifths (musical chords)" and so on:

```
multiplication using, 1093         and generalizing numbers, 1158     Isotropy in, 980
recursive algorithm for, 1142        universality of, 1159            methods based on, 940
see also Fourier                     see also Finite fields           neighborhood compared to
Fiber bundles                      Field theory (physics)                cellular automata, 928
  and continuum limits of            history of, 1024                 for PDEs, 924
    networks, 1030                   nonlinear PDE as, 923             and reaction-diffusion, 1013
  and gauge theories, 1045           quantum, 1061                    Finite element methods, 940
Fiber optics, vs. broadcasting, 1188 see also Quantum field theory    Finite fields
Fibers (biological)                Fields (agricultural)                as not universal, 1160
  and folding of tissue, 417         patterns of from space, 1187      see also Additive cellular automata
Fibonacci, Leonardo (Pisano) (Italy, Fifths (musical chords)           see also Field theory (in abstract
  ~1170 - ~1250)                     curves of, 146                      algebra)
  and digital numbers, 902           perfect, 1079                    Finite groups
  and Fibonacci numbers, 891       Fig leaves, 1005                     axioms for, 1176
  and rabbit populations, 1002     Figurate numbers, 911                Cayley graphs of, 1032
  and tables of primes, 910        Filters (for data)                   as extraterrestrial messages, 1190
  and trees, 893                     cellular automata as, 225          rules for, 938
Fibonacci (Fibonacci sequence)       in visual perception, 1076       Finite impulse response
  difficulty of making with CAs,   Filters in posets, 1040               and sequential CAs, 1035
    1186                           Final cause, 1185                  Finite-size scaling
  and entropy in rule 32, 958      Final theory                         in Ising models, 983
  generalized, 891                   see Ultimate theory of physics   Finite-size systems, 255-260, 961
  generalized for randomness      Financial systems                   Finite state machines
    generation, 975                  applications to, 429-432           see Finite automata
  generating function for, 1091     data from as source of           Finitely presented groups
  as initial condition for rule 60,    randomness, 969                  see Groups
    1091                             history of models of, 1015      Finitistic mathematics, 1158
  leading digits in, 914             meaning of random data in, 1183 Finkelstein, David R. (USA, 1929- )
  and multiway system states, 205   simulations of, 968                and discreteness of space, 1027
  and multiway systems based on     using randomness to verify         in Preface, xiii
    numbers, 939                      contracts in, 968              Finnish, logic operations in, 1173
  number with digits at, 914, 1070 FindMinimum                       Fins (heat exchanger)
  in ordering of math constructs,    and network layouts, 1031         characteristic shapes of, 1183
    1177                           FindRoot                           Fire
  and plant phyllotaxis, 1006        difficulty of evaluating, 1134    as artifact, 1183
  and polyominoes, 943               iterative algorithm for, 1141    as basis for universe, 1125
  as precursors to my work, 878   Fine structure constant ($\alpha$)   as visible from space, 1187
  properties of, 890                 numerology for, 1025            Firing of neurons
  and prosody, 875                   and perturbation theory, 1057     in neural networks, 1102
  and randomness generators, 975  Fine tuning                           repeatable randomness in, 976
  and recursion history, 907        and self-organized criticality, 989 Firing squad problem (in cellular
  as recursive sequence, 128      Fingerprints                           automata), 876, 1035
  and rule 150 pattern, 885         origin of patterns in, 1013      First digits, 914
  as solution to Diophantine        randomness in, 1014                of powers, 903
    equation, 1161                Fingers, formation of human, 419   First Law of Thermodynamics, 1019
  and spectral maxima, 1081       Finite automata                    First-order phase transitions, 981
  and substitution systems, 82, 890 and attractors for CAs, 277      Fisher, Ronald A. (England,
  as term in continued fraction, 913 and Boolean functions, 1097       1890-1962)
                                     and CA encodings, 1119            and random number tables, 968
```

In the end the index—even printed as it was in 4 columns—ran to 80 pages (or more than 6% of the book). It was obviously a very useful index, and it could even be entertaining to read, not only for its eclectic jumps from one term to the next, but also for the unexpected terms that appeared. What's "Flash photography" or "Flint arrowheads" doing there, or "Frogs" for that matter? What do these terms have to do with a new kind of science?

But for all its value, I was a bit concerned that the index might be so long that it finally made the book "too long". Even without the index the book ran to 1197 pages. But why tell people, I thought, that the whole book is really 1280 pages, including the index? If the pages of the index were numbered, then one could immediately see the number of that last page. But why number the pages of an index? Nobody needs to refer to those pages by numbers; if anything, just use the alphabetized terms. So I decided just quietly to omit the page numbers of the index, so we could report the length of the book as 1192 pages.

## How to Publish a Book

OK, so *A New Kind of Science* was going to be a book. But how was it going to be published? At the time I started writing *A New Kind of Science* in 1991 the second edition of *The Mathematica Book* had just been released, and its publisher (Addison-Wesley) seemed to be doing a good job with it. So it was natural to start talking about my new book with the same publisher. I was quite aware that Addison-Wesley was primarily a publisher of textbook-like books, and in fact the particular division of Addison-Wesley that had published *The Mathematica Book* was more oriented towards monographs and special projects. But the success of *The Mathematica Book* generated what seemed like good corporate interest in trying to publish my new book.

But how would the details work? There were immediate questions even about printing the book. I knew the book would rely heavily on graphics which would need to be printed well. But to print them how they needed to be printed was expensive. So how would that work financially? (And at that point I didn't yet even know that the book would also be more than a thousand pages long.)

The basic business model of publishing tends to be: invest up front in making a book, then (hopefully) make money by selling the book. And for most authors, the book can't happen without that up-front investment. But that wasn't my situation. I didn't need an advance to support myself while writing the book. I didn't need someone to pay for the production of the book. And if necessary I could even make the investment myself to print the books. But what I thought I needed from a publisher was access to distribution channels. I needed someone to actually sell books to bookstores. I needed there to be a sales team that had relationships with bookstore chains, and that would do things like actually visit bookstores and get books into them.

And in fact quite a lot of the early discussion about the publishing of the book centered around how salespeople would present it. How would the book be positioned relative to the well-known "popular science" books of the time? (That positioning would be key to the size of initial purchases bookstores might make.) What special ways might the salespeople make the book memorable? Could we get enough textile cone shells that the salespeople could drop one off at every bookstore they visited? (The answer, it was determined, was yes: in the Philippines such shells were quite plentiful.)

But how exactly would the numbers work? Bookstores took a huge cut (often above 50%). And if the book was expensive to print, that didn't leave much of a margin. At least at the time, the publishing industry was very much based on formulas. If you spend $x to print a book, you need to spend $y on marketing, and you pay the author $y (yes, same y) as an advance on royalties. For the author, the advance serves as a kind of guarantee of the publisher's effort—since unless the book sells, the publisher just loses that money.

Well, I most definitely wanted a guarantee that the publisher would put effort in. But I didn't need or want an advance; I just wanted the publisher to put as much as possible into distribution. Around and around it went, trying to see how that might work. Exasperated, I found an expert on book deals. They didn't seem to be able to figure it out either. And I began to think: perhaps I should go to a different publisher, maybe one more familiar with widely distributed books.

It's typical for authors not to interact directly with such publishers, but instead to go through an agent. In principle that allows authors not to have to exercise business savvy, and publishers not to be exposed to the foibles of authors. But I just wanted to make what—at least by tech industry standards—was a very simple deal. One agent I'd known for a while insisted that the key was to maximize the advance: "If the book earns out its advance [i.e., brings in more royalties from actual sales than were paid out up front], I haven't done my job." But that wasn't my way of doing business. I wanted both sides in any deal to do well.

Then there was the question of which publisher would be the right one. "Sell to the highest bidder", was the typical advice. But what I cared about was successful book distribution, not how much a publisher might (perhaps foolishly) spend to get the book. Particularly at the time, it was a very clubby but strangely dysfunctional industry, full of belief in a kind of magic touch, but also full of stories of confusion and failure. Still, I thought that access to distribution channels was important enough to be worth navigating this.

And by 1993 quite a bit of time had been spent on discussions about publishing the book. A particular, prominent New York publisher had been identified, and the process of negotiating a contract with them was underway. From a tech industry point of view it all seemed quite Victorian. It started from a printed (as in, on a printing press) 70-page contract that seemed to date from 20 years earlier. Though after not very long, essentially every single clause had been crossed out, and replaced by something different.

An effort to "show what value they could bring" led to the incident about cover designs mentioned above. And then there was the story about printing, and printing costs. The terms of our potential deal made it quite important to know just how much it would cost to print the book. So to get a sense of that we got quotes from some of our usual printing vendors (and, yes, in those days before the web, a software company like ours did lots of printing). The publisher insisted that our quotes were too high—and that they could print the book much more cheaply. My team was skeptical. But at the center of this discussion was an important technical issue about how the book would actually be printed.

Most widely distributed ("trade") books are printed on so-called web presses—which are giant industrial machines that take paper from a roll and move it through at perhaps 30 mph. (The term "web" here refers to the "web of paper" on its path through the machine, not the subsequently invented World Wide Web.) A web press is a good way to print a just-read-the-words kind of book. But it doesn't give one much control for pictures; if everything's running through at high speed one can't, for example, carefully inject more ink to deal with a big area of black on a specific page.

And so if one wanted to print a more "art-quality" book one had to use a different approach: a sheet-fed press in which each collection of pages is "manually" set up to be printed separately on a large sheet of paper. Sheet-fed presses give one much more control—but they're more expensive to operate. The printing quotes we'd got were for sheet-fed presses, because that was the only way we could see printing the book at the quality level we wanted. (I was sufficiently curious about the whole process that I went to watch a print run for something we were printing. In interacting with our potential publisher, I was rather disappointed to discover that none of the editorial team appeared to have ever actually seen anything being printed.)

But in any case the publisher was claiming that they knew better than us, and that they could get the quality we needed on a web press, at a much lower price. They offered to run a test to prove it. We were again skeptical: to do the setup for a web press is an expensive process, and it makes no sense to do it for anything other than a real print run of thousands of books. But the publisher insisted they could do it. And our only admonition was "Don't show us a result claiming it was made on a web press when it wasn't!".

A few weeks went by. Back came the test. "You can't be serious", we said. "That's a sheet from a sheet-fed press; we can see the characteristic registration marks!" I never quite figured out if they thought they could pull the wool over our eyes, or if this was just pure cluelessness. But for me it was basically the last straw. They came back and said "Why don't we just refactor the contract and give you a really big advance?"

"Nope", I said "you're profoundly missing the point! We're done." And that's how—in 1995—we came to make the decision to publish *A New Kind of Science* "ourselves".

But when I say "ourselves" there was quite a bit more to that story. Back at the beginning of 1995 we were thinking about the upcoming third edition of *The Mathematica Book*, and realizing that we needed to re-jigger its publishing arrangements. And while the machinations with publishers about the NKS book had been a huge waste of time, they had helped me understand more about the publishing industry—and made me decide it was time for us to create our own publishing "imprint", Wolfram Media.

Its website from 1996 (I never liked that logo!) highlights our first title—the co-published third edition of *The Mathematica Book*:

This was soon joined by other titles, like our heavily illustrated *Graphica* books. But it wasn't until 1999 that I began to think more seriously about the final publishing of the NKS book. In the fall of 1999 we duly listed the book with the large bookstore chains and book distributors, as well as with the already-very-successful Amazon.

And in late 2000 we started touting the book on our now-more-attractive website as "A major release coming soon...":

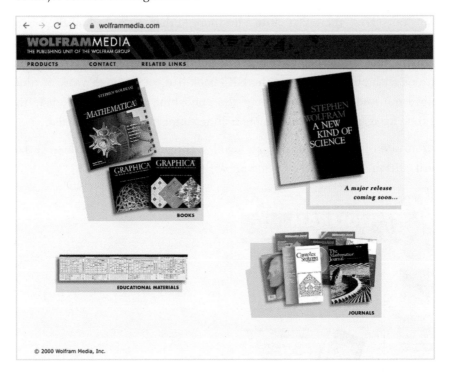

Particularly in those days, the typical view was that most of the sales of a book would happen in the first few weeks after it was published. But—as we'll discuss later—printing a book (and especially one like the NKS book) takes many weeks. So that creates a tricky situation, in which a publisher has to make a high-stakes decision about how many books to print at the beginning. Print too few books and at least for a time, you won't be able to fill orders, and you'll lose out on the initial sales peak. Print too many books and you'll be left with an inventory of unsold books—though the more books you print in a single print run, the more you'll spread the initial setup cost over more books, and the lower the cost of each individual book will be.

Bookstores were also an important part of the picture. Books were at the time still predominantly bought through people physically browsing at bookstores. So the more copies of a book a bookstore had, the more likely it was that someone would see it there, and buy it. And all this added up to a big focus of publishing being on the size of the initial orders that bookstores made.

How was that determined? Mostly it was up to the buyers at bookstores and bookstore chains: they had to understand enough about a book to make an accurate prediction of how many they'd be able to sell. There was a complicated dance through which publishers signaled their expectations, saying for example "X copy initial print run", "X-city promotional tour", "$X promotional budget". But in the end it was a very person-to-person sales process, often done by traveling-around-the-country salespeople who'd developed relationships with book buyers over the course of many years.

How were we going to handle this? It certainly helped that by late 2000 there were starting to be lengthy news articles anticipating the book. And it also helped that one could see that the book was gaining momentum on Amazon. But would a sales manager we had who was used to selling software be able to sell books? At least in this case the answer was yes, and by the end of 2001 there were starting to be substantial orders from bookstores.

By the time I finished writing the book at the beginning of 2002 we were in full "book-publishing" mode. There were still lots of issues to resolve. How would we handle distribution outside the US? (We'd actually had a UK co-publisher lined up but we eventually gave up on them.) How would we reach the full range of independent bookstores? And so on. Looking at my archives I find mail from April 2002 in which I was contacting Jeff Bezos about a practical issue with Amazon; Jeff responded that he "couldn't wait to read [the book]", noting that "For a serious book like yours, we often account for a substantial fraction of sales." He was right—and in fact the NKS book would reach the #1 bestseller slot on Amazon.

By the beginning of 2002 we'd had a design for the front cover of the NKS book for six years. But what about the back cover? It's traditional to put quotes ("blurbs") on the backs of books that people will browse in bookstores. So, in February 2002 we sent a few draft copies of the book to people we thought might give us interesting quotes. Probably the most charming response was Arthur C. Clarke's report of the delivery of the book to his house in Sri Lanka:

> **Date:** Fri, 01 Mar 2002 16:46:44 +0600
> **From:** "Sir Arthur C. Clarke" <blenheim@sri.lanka.net>
> **To:** "Stephen Wolfram" <s.wolfram@wolfram.com>
>
> Dear Stephen,
>
> A ruptured postman has just staggered away from my front door...
>
> Stay tuned.....
>
> Arthur 1 Mar 2002

A few days later, he emailed again "Well, I have <looked> at (almost) every page and am still in a state of shock. Even with computers, I don't see how you could have done it-", offering the quote "Stephen's magnum opus may be the book of the decade, if not the century", then adding "Even those who skip the 1200 pages of (extremely lucid) text will find the computer-generated illustrations fascinating. My friend HAL is very sorry he hadn't thought of them first..."

Other quotes came in too. At his request, I'd sent Steve Jobs a copy of the book—and I asked if he'd like to provide a quote. He responded that he thought I really shouldn't have quotes on the back of the book. "Isaac Newton didn't have quotes; nor should you." And, yes, Steve had a point. I was trying to write a book that would have long-term value; it didn't really make sense to have moment-of-publication quotes printed on it.

So—feeling bad for having solicited quotes in the first place—we dropped them from the back cover, instead just putting images from the book that we thought would intrigue people:

Still, my team did use Arthur C. Clarke's quote on the publishing-industry-obligatory ad we ran in *Publisher's Weekly* on April 15 as part of a final sprint to increase up-front orders from bookstores:

At least the way the book trade was in those days, there was a whole arcane dance to be done in publishing a book—with carefully orchestrated timing of book reviews, marketing initiatives at bookstores, and so on. My archives contain a whole variety of pieces related to that (many of which I don't think I saw at the time). One of the more curious (whose purpose I don't now know) involves a perhaps-not-naturally-colored lizard that could be viewed as having escaped from page 426 of the book:

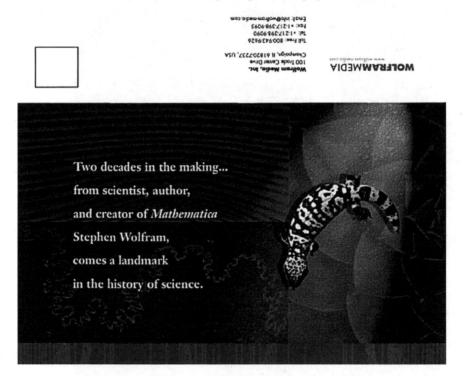

## How Are We Going to Print the Book?

From the very beginning I was very committed to doing the best we could in actually printing the book. My original discoveries about rule 30 and its complexity had originally crystallized back in 1984 when I'd first been able to produce a high-resolution image of its behavior on a laser printer. Book printing allowed still vastly higher resolution, and I wanted to make use of that to make the NKS book serve if nothing else as a "printed testament" to the idea that complexity can be generated from simple computational rules.

The Making of *A New Kind of Science*

Here's what a printout of rule 30 made on a laser printer looks like under a microscope (this printout is from 1999, but it basically looks the same from a typical black-and-white laser printer today):

And here's what the highest-resolution picture of rule 30 from the printed NKS book looks like (and, yes, coincidentally that picture occurs on page 30 of the book):

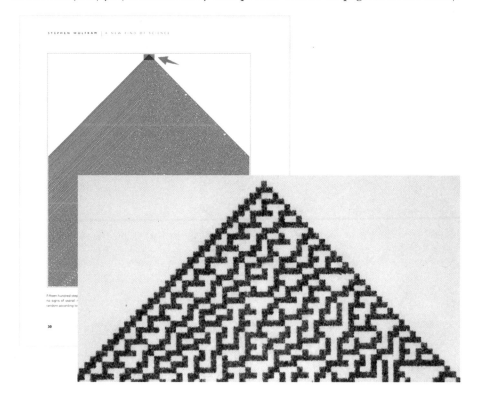

197

You can see the grain of the paper, but you can also see crisp boundaries around each cell. To give a sense of scale, here's a word from the text of the book, shown at the same magnification:

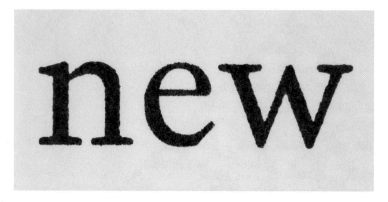

To achieve the kind of crispness we see in the rule 30 picture (while, for example, keeping the book of manageable size and weight) was quite an adventure in printing technology. But the difficulties with pure black and white (as in this picture of rule 30) paled in comparison to those involved with gray scales.

The fundamental technology of printing is quite binary: there's either ink at a particular place on a page, or there isn't. But there's a standard method for achieving the appearance of gray, which is to use halftoning, based essentially on an array of dots of different sizes. Here's an example of that from the photograph of a tiger on page 426 of the NKS book:

But one feature of photographs is that they mostly involve smooth gradations of gray. In the NKS book, however, there are lots of cases where there are tiny cells with different gray levels right next to each other.

Here's one example (from page 157—which we'll encounter again later):

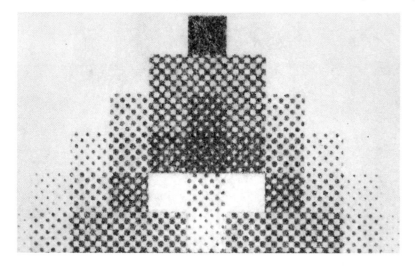

Here's another example with slightly smaller cells (page 640):

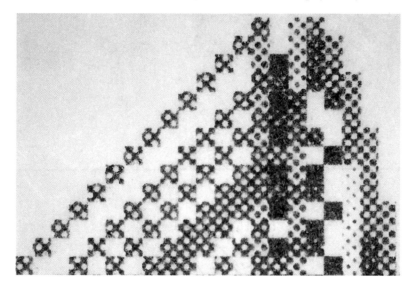

Here's a nice example based from a 3D graphic (page 180):

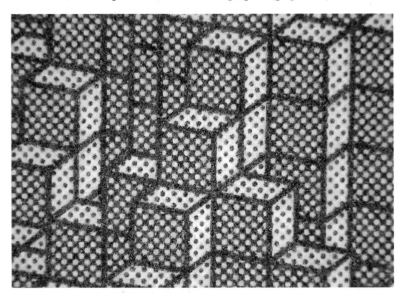

And here's one where the gray cells are so small that the halftoning gets mixed up with the actual boundaries of cells (page 67):

But in general to achieve well-delineated patches of gray there have to be a decent number of halftone dots inside each patch. And this is one place where we were pushing the boundaries of printing technology for the NKS book. Here's an image from a 1995 print test (and, yes, we were testing printing as early as 1992):

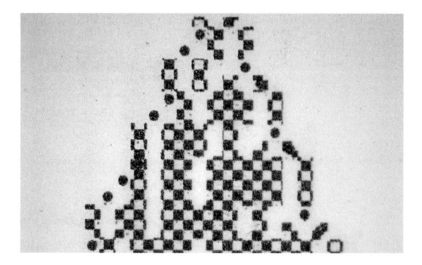

This is a more straightforward case, because we're dealing with exactly 50% gray. But look at the difference for the same picture in the final NKS book:

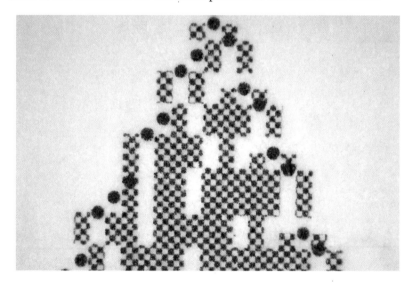

We slightly changed our standard for how big the mobile-automaton-active-cell dots should be. But the main thing to notice is that the halftone checkerboard in each gray cell is roughly twice as fine in the final version. In printing terminology, the 1995 test used a standard "100-line screen"; the final NKS book used a "175-line screen" (i.e. basically 175 dots per inch).

The importance of this is even more obvious when we start looking not just at gray cells, but also at gray lines. Here's the 100-line-screen print test:

And here's the same picture in the final book:

Here's the picture that first introduces rule 30:

And a big issue was: how thin can the gray lines be, while not filling in, and while still looking gray? That was a difficult question, and was only answered by lots of print testing. One of the main points was: even if you effectively specify dots of a certain size, what will be the actual sizes of dots formed when the ink is absorbed into the paper? And similarly: will the ink from black cells spread into the area of the gray line you're trying to print between them? In printing it's typical to talk about "dot gain". If you think you're setting up dots to give a certain gray level, what will be the actual gray level you'll get when those dots are made of ink on paper?

We were constantly testing things like this, with different printing technology, different paper and so on:

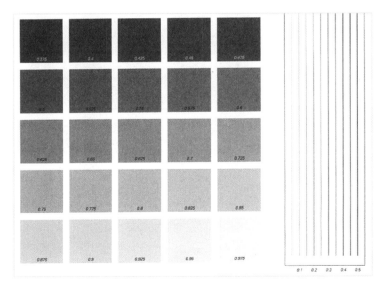

We used a "densitometer" (yes, this was before modern digital cameras) to measure the actual gray level, and deduce the dot gain function. And we tested things like how thin lines could be before they wouldn't print.

In halftoning, one effectively applies a global "screen" (as in, something with an array of holes in it, just like in pre-digital printing) to determine the positions of dots. We considered effectively setting up our own dot placement algorithm, that would for example better align with cells in something like a cellular automaton. But tests didn't show particularly good behavior, and we soon reverted to considering the "traditional approach", though with various kinds of tweaking.

Should the halftone dots be round, or elliptical? What should the angle of the array of dots be (it definitely needed to avoid horizontal and vertical directions)? As this manifest indicates, we did many tests:

| Letter | Page Number | Screen | Dot | Angle |
|---|---|---|---|---|
| A | 246 | 200 | elliptical | 45° |
| B | 637 | 200 | round | 75° |
| C | 246 | 200 | round | 75° |
| D | 082 | 175 | round | 75° |
| E | 224 | 200 | round | 45° |
| F | 336 | all type | | |
| G | 082 | 200 | round | 45° |
| H | 224 | 200 | round | 75° |
| I | 335 | no halftones | | |
| J | 637 | 200 | round | 45° |
| K | 637 | 175 | round | 45° |
| L | 082 | 200 | elliptical | 45° |
| M | 637 | 200 | elliptical | 45° |
| N | 637 | 200 | round | 75° |
| O | 246 | 200 | elliptical | 75° |
| P | 224 | 200 | elliptical | 45° |
| Q | 335 | no halftones | round | |
| R | 335 | no halftones | round | |
| S | 335 | no halftones | round | |
| T | 246 | 200 | elliptical | 45° |
| U | 335 | no halftones | round | |
| V | 335 | no halftones | round | |
| W | 082 | 200 | elliptical | 75° |

The final conclusion was: round dots, 175-line screen, 45° angle. But it took quite a while to get there.

But, OK, so we had a pipeline that started with Wolfram Language code, and eventually generated PostScript. Most of the complexity we've just been discussing came in converting that PostScript to the image that would actually be printed. And in imaging technology jargon, that's achieved by a RIP, or raster image processor, that takes the PostScript and generates a bitmap (normally represented as a TIFF) at an appropriate resolution for whatever will finally render it.

In the 1990s the standard thing to do was first to render the bitmap as a negative onto film. And my archives have tests of this that we did in 1992, here again shown under a microscope:

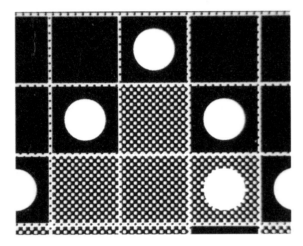

Everything looks perfectly clean. And indeed printing this purely photographically still gives a perfectly clean result:

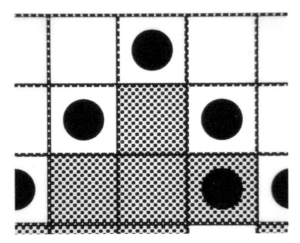

But it gets much more complicated when one actually prints this with ink on a printing press:

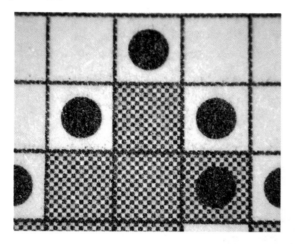

The basic way the printing is done is to ("lithographically") etch a printing plate which will then be inked and pressed onto paper to print each copy. Given that one already has film, one can make the plate essentially photographically—more or less the same way microprocessor layouts and many other things are made. But by the beginning of the 2000s, there was a new technology: direct-to-plate printing, in which an (ultraviolet) laser directly etches the plate (a kind of much-higher-resolution "plate analog" of what a laser printer does). And in order to get the very crispest results, direct-to-plate printing was what we used for the NKS book.

What's the actual setup for printing? In the sheet-fed approach that we were using, one combines multiple pages (in our case 8) as a "signature" to be printed from a single plate onto a single piece of paper. Here's a (yes, rather-unremarkable-looking) actual plate that was used for the first printing of the NKS book:

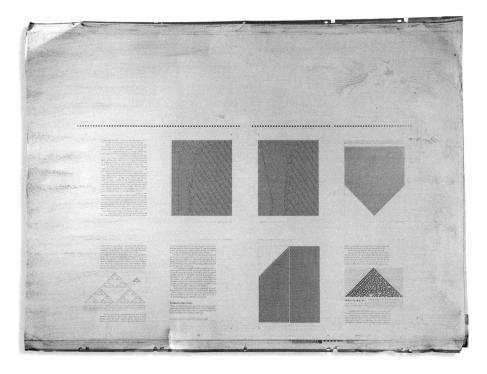

And here's an example of a signature printed from it, with pages that will subsequently be cut and folded:

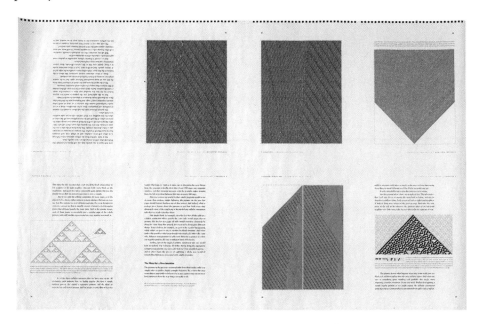

Under a microscope, the plate looks pretty much like what will finally be printed onto the paper:

But now the next big issue is: what kind of paper should one use? If the paper is glossy, ink won't spread on it, and it's easier to get things crisp. But adding a glossy coating to paper makes the paper heavier and thicker, and we quickly determined that it wasn't going to be practical to print the NKS book on glossy paper. Back in the 1980s it had become quite popular to print books on paper that looked good at first, but after a few years would turn yellow and disintegrate. And to avoid that, we knew we needed acid-free paper.

Any particular kind of paper will come in different "weights", or thicknesses. And the thicker the paper is, the more opaque it will be, and the less see-through the pages of the book will be—but also the thicker the book will be with a given number of pages. At the beginning we didn't know how long the NKS book would be, and we were looking at comparatively thick papers; by the end we were trying to use paper that was as thin as possible.

Back in 1993 we'd identified Finch Opaque as a possible type of paper. In 1995 our paper rep suggested as an alternative Finch VHF ("Very High Finish")—which was very smooth, and was quite bright white. But normally this paper was used in very thick pages. Still, it was possible for the paper mill to produce thinner versions as well. We studied the possibilities, and eventually decided that a 50-lb version (i.e. with the paper weighing 50 lbs per 500 uncut sheets) would be the best compromise between bulk and opacity. So 50-lb Finch VHF paper is what the NKS book is printed on.

Paper, of course, is made from trees. And as I'll explain below, during the publishing of the NKS book, I became quite aware of the physical location of the trees from which the paper for the NKS book was made: they were in upstate New York (in the Adirondacks). At the time, though, I didn't know more details about the trees. But a few years ago I learned that they were eastern hemlock trees. And it turns out that these coniferous trees are unusual in having long fibers—which is what allows the paper to be as smooth as it is. Talking about hemlock makes one think of Socrates. But no, hemlock the poison comes from the "poison hemlock" plant (*Conium maculatum*), which is unrelated to hemlock trees (which didn't grow in Europe and seem to have gotten their hemlock name only fairly recently, and for rather tenuous reasons). So, no, the NKS book is not poisonous!

Once signatures are printed, the next thing is that the signatures have to be folded and cut—in the end forming little booklet-like objects. And then comes the final step: binding these pieces together into the finished book. By the mid-1990s *The Mathematica Book* had given us quite a bit of experience with the binding of "big books"— and it wasn't good. Many copies of multiple versions of *The Mathematica Book* (yes, not printed by us) had basically self-destructed in the hands of customers.

How were we going to be sure this wouldn't happen for the NKS book? First, many books—including some versions of *The Mathematica Book*—were basically "bound" by just gluing the signatures into the "case" of the book (with little fake threads added at the ends, for effect). But to robustly bind a big book one really has to actually sew the signatures to the case, and a standard way to do this is what's called Smythe sewing. And that's what we determined to use for the NKS book.

Still, we wanted to test things. So we sent books to a book-testing lab, where the books were "tumbled" inside a steel container, 1200 times per hour, "impacting the tail, binding edge, head and face" of each book 4800 times per hour. After 1 hour, the lab reported "spine tight and intact". After 2 hours "text block detached from cover". But that's basically only after doing the equivalent of dropping the book thousands of times!

As we approached the final printing of the NKS book, there were other decisions to be made. The endpapers were going to have a rule 30 pattern printed on them. But what color should they be? We considered several, picking the goldenrod in the end (and somehow that color now seems to have become the standard for the endpapers of all books I write):

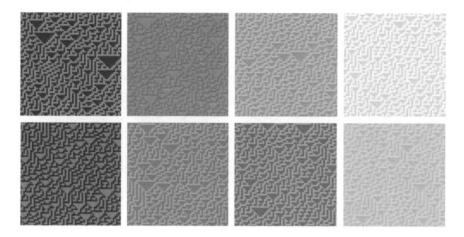

In the late stages of writing the NKS book one of the big concerns was just how long the book would eventually be. We'd figured out the paper, the binding, and so on. And there was one hard constraint: the binding machines that we were going to use could only bind a book up to a certain thickness. With our specs the limit was 80 signatures—or 1280 pages. The main text clocked in at 1197 pages; with front matter, etc. that was 1213 pages. But then there was the index. And I was writing a very extensive index, that threatened to overrun our absolute maximum page count. We formatted the index in 4 columns as small and tight as we thought we could. And in the end it came in just under the wire: the book was 1280 pages, with not a single page to spare. (Somewhat simplifying the story, I've sometimes said that after a decade of work on the NKS book, I had to stop because otherwise I was going to have a book that was too long to bind!)

**The Great Printing Adventure**

High-quality printing of the kind needed for the NKS book was then—and is now—often done in the Far East. But anticipating that we might need to reprint the book fairly quickly we didn't consider that an option; it would just take too long to transport books by boat across the Pacific. And conveniently enough, we determined that there was a cost-effective North American alternative: print the book in Canada. And so it was that we chose a printer in Winnipeg, Canada, to print the NKS book.

On February 7, 2002, the files for the book (which were now PDF, not pure PostScript) were transferred (via FTP) to the printer's computers—a process which took a mere 90 minutes. (Well, it had to be done twice, because of an initial glitch.) But then the next step was to produce "proofs" for the book. In traditional printing, where printing plates were made from film, one could produce the film first, then make a photographic print of this, check it, and only then make the plates. But we were going to be making plates directly. So for us, "proofing" was a more digital process, that involved using a separate device from the one that would actually make the plates. Supposedly, though, "the bits were the bits", and the results would be the same.

Within a couple of days, the printer had the first proofs made, and a few issues were seen—such as white labels inside black cells simply disappearing. The cause was subtle, though didn't take a long time to find. Some 3D graphics in the book had generated color PostScript—and in all our tests so far these had just automatically been converted to grayscale. But now the presence of color primitives had made the RIP that was converting from PostScript change its settings—and cause other problems. But soon that was worked around, and generating proofs continued.

By February 14 we had the first batch of proofs in our hands, and my team and I went to work going through them. Everything looked just fine until—ugh—page 157:

That was supposed to be a symmetrical (continuous) cellular automaton! So how could it be different on the two sides? Looking now under a microscope, here are the corresponding places on the two sides:

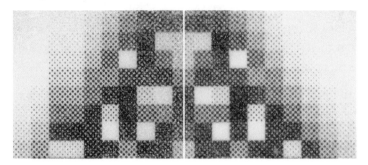

And we can see that somehow on the left an extra column of cells has mysteriously appeared. But where did it come from? We checked the original PostScript. Nope, it wasn't there. We asked the printer to rerun the proof, and, second time around, it was gone. Very mysterious. But we figured we could go ahead—and in any case we had a tight schedule to meet.

So on February 17 the book designer who'd worked on the project ever since the beginning went to Winnipeg, and on February 18 the book began to be printed.

I wasn't there (and actually now I wish I'd gone) but a bunch of pictures were taken. After a decade of work all those abstract bits I'd produced were being turned into an actual, physical book. And that took actual industrial work, with actual industrial machines:

Here's the actual press that's about to print a signature of the NKS book (the four "stations" here are set up to print four different colors, but we were only using one of them):

And here's that signature "coming off the press":

It really was coming out "hot off the press"—with a machine drying off the ink:

Those controls let one change ink flows and pressures to make all the pages come out correctly balanced:

Thanks, guys, for checking so carefully:

Pretty soon there were starting to be lots of copies of signatures being printed:

And—after being involved for more than a decade—the book designer was finally able to sign off on the printed version of the opening signature of the book:

The whole process of printing all the signatures of the book was scheduled to take about four weeks. We had been receiving and checking the signatures as they were ready—and on March 12 we received the final batch, and began to check them, on the alert for any possible repeat of something like the page-157 problem.

Within a few hours a member of our team got to page 332 (on "signature 21") which included this image:

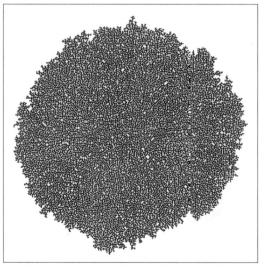

(a)

I'm frankly amazed he noticed, but if you look carefully near the right-hand edge you might be able to tell that there's a strange kind of "seam". Zoom in at the top and you'll see:

And, yes, this is definitely wrong: with the aggregation rule used to make this picture it simply isn't possible to have floating pieces. In this case, the correct version is:

An hour or so later two more glitches were found, on pages 251 and 253. Both cases again involved something like a column of cells being repeated. On page 253 zooming into the image

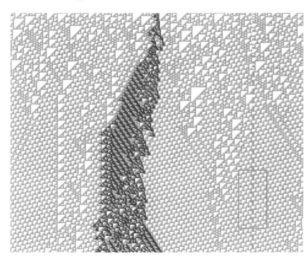

reveals strange and "impossible" imperfections in the supposedly periodic background of rule 110:

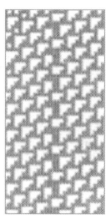

On page 194 there was another glitch: an arrow on a graph that had basically become too thin to see. But this problem at least we could understand—and it was our fault. Instead of setting the thickness of the arrow in some absolute way, we'd just set it to be "1 pixel"—which in the final printing was too thin to see.

But what about the other glitches? What were they? And might there be more of them?

The signatures from the book were ready to start being bound. Should we hold off and reprint the signatures where we'd found glitches? Could we do this without blowing our (already very tight) schedule? Could we even get enough extra paper in time? My team was adamant that we should try to fix the glitches, saying that otherwise they would "nag at us forever". But I wanted first to see if we could characterize the bug better.

We knew it was associated with the rendering of the PostScript image operator. Even though PostScript is basically a vector graphics description language, the image operator allows one to include bitmaps. Normally these bitmaps are used to represent things like photographs, and have tiny ("few-pixel") cells. But in the cellular-automaton-like images we were having trouble with, the cells were much larger; in the case of page 157, for example, each one was roughly 75 of the final 2400-dpi pixels across. This was absolutely something the image operator was set up to handle. But somehow something was going wrong.

And what was particularly surprising is that it seemed as if the problem was happening after the PostScript was converted to a TIFF. Could it perhaps be in the driver for both the proofing and the final plate production system? Time was short, and we needed to make a decision about what to do.

I fired off an email to the CEO of the company that made the direct-to-plate system, saying: "We of course do not know the details of your software and hardware systems. However, we have done a little investigation. It appears that the data ... in the case of this image is a bilevel TIFF with LZW compression. We speculate that the LZW dictionary contains something close to the actual squares seen in the image, and that somehow pointers to dictionary entries are being corrupted or are not being used correctly in the decompression of the TIFF. The TIFF experts at my company say they have never seen anything like this in developing software based on standard imaging libraries, making us suspect that it may be some kind of buffering or motion optimization bug associated with your actual hardware driver."

The CEO of what was by then quite a large company had personally designed the original hardware, and when we talked by phone he speculated that what we were seeing might be some kind of obscure mechanical issue with the hardware. But his chief of software soon sent mail explaining that "of the several hundred thousand books that go through [their system] each year, there are a couple that have imaging problems like this." But, he added, "Usually they are books about halftone screening algorithms, which cause an almost-recursive problem...". He said the specific issue we were having looked like a "difficult to reproduce problem we have known about for some time but is transient enough that re-imaging the same file can 'correct' the problem." He added that: "Our hypothesis is that it is related to a memory access error in the RIP that manifests only at low-memory conditions, or after many allocation/deallocation cycles of RAM blocks. The particular code path is not one we have source-code access to, and is rumored to be many years old, so not many people on earth are prepared to make substantive changes to it."

OK, so what next? The RIP had been developed by Adobe, creators of PostScript. So I emailed John Warnock, co-founder of Adobe, who I'd met at quite a few software-industry get-togethers before my NKS-book "hermit period". I commented that "One thing that's peculiar (at least without knowing how the RIP works) is that the glitch involves overwriting of a column ... even though scanning the underlying PostScript would involve going from one row to the next." Warnock responded helpfully, copying his team, though saying (in an echo of what we'd already heard) "I don't know who does PostScript stuff anymore".

Well, that seemed like pretty much the end of the road. So we decided to assume that the glitches we'd found were the only ones, and—for perfection's sake—we'd reprint those signatures, which by that point the printer had helpfully said they could do without blowing the schedule.

Two weeks later, Adobe delivered a new version of the RIP, in which they believed the bug had been fixed, noting that there had been significant code cleanup, and they were now using a newer version of the C++ compiler. Meanwhile, I'd realized another issue: a variety of magazines had requested files from us to be able to print high-resolution images from the book. Would they end up using the same software pipeline, and potentially have the same problem? A general release of any fix was still quite far away.

Meanwhile, with the two "glitch" signatures reprinted, the book was off to be bound. The cover had also been printed, now making use of all four stations of the presses. Under a microscope the characteristic "rosettes" of 4-color printing are visible:

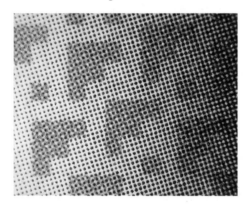

Actually, the book in a sense has two covers: a detachable dust jacket (including a dated picture of me!) and a "permanent" hard cover—which I think looks very nice:

But as I was just now looking back through my archives I found an email from February 2002, expressing concerns about the fading of ink on the cover. The printer assured us that we had "nothing to worry about unless the books were exposed to direct sunlight for an extended amount of time." But then they added "The reds and yellows will fade faster than the other pigments, but this is not something that would be noticeable in the first 20–40 years." Well, it's now been 20 years, and it so happens that I have a copy of the NKS book that's been exposed to sunlight for much of that time—and look what's happened to its spine, right on cue:

I received a first, hand-bound, finished NKS book on April 22. And very soon books were on their way to bookstores and distribution centers. And people were ordering the book—in large numbers. And that meant that the books we'd printed so far weren't going to be enough. And on May 12—two days before the May 14 official publication date of the book—another print run was started.

Fortunately it was possible to reuse the plates from the first print run (well, apart from the one which said "First printing"), so we didn't have to worry about new glitches showing up.

But once the book was published, demand continued to be strong, and on June 4 we needed to do another print run. And this time new plates had to be made. Were there going to be new glitches? We decided we should check the plates before we started printing—so we sent the person who'd caught the glitches before on a trip to Canada. Turns out the bug hadn't yet been fixed, and there it was again on pages 583 and 979.

Some time later I heard that the bug was finally found and fixed, and had been lurking in the implementation of the PostScript image operator for well over a decade. Yes, software is hard. And computational irreducibility is rampant. But in the years since the NKS book was published, no other weird glitches like this have ever shown up. Or at least nobody has ever told us about any.

But as I was writing this, I wondered: what became of that other glitch that was in the first printing—the one with the thin arrows that was our fault? I opened an NKS book from my desk. No problem. But then I pulled off my shelf the leather-bound copy of the first printing that my team made for me, and turned to page 194. And there it was—the "1-pixel arrow" (compared here under a microscope to the second printing):

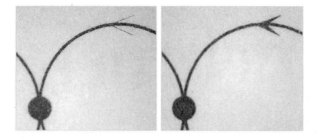

And yet one more thing: looking in my archives, I find a cover sheet for a print test from March 1, 1999—which notes that there is "glitch with the graphic on page 246" ... "which has been traced to a problem with the Adobe 4.1 PostScript driver" for the RIP—made by a completely different company:

Was it the same "page-157" bug? I looked for the print test. And there's "page 246" (which ended up in the final version as page 212):

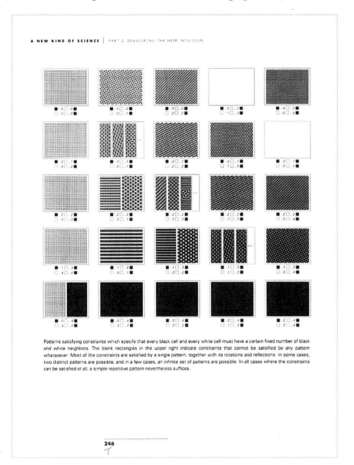

Under a microscope, most of the arrays of cells look just fine:

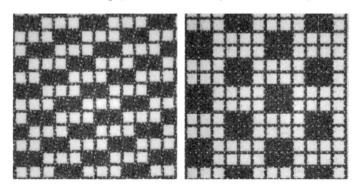

But there it is: something weird again!

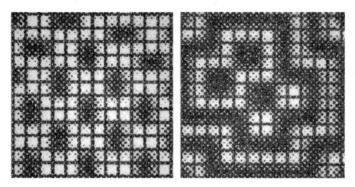

Is it the same "page-157" bug? Or is it another bug, perhaps even still there, 23 years later?

## The Great Printing Adventure, Part 2

When the NKS book was officially published on May 14, 2002, it was the #1 bestselling book on Amazon, and it was steadily climbing the *New York Times* and other bestseller lists. We'd just initiated a second printing, which would be finished in a few weeks. But based on apparent demand that printing wasn't going to be sufficient. And in fact a single bookstore chain had just offered to buy the whole second printing. We initiated a third printing on June 4, and then a fourth on June 18. But if we were going to keep the momentum of sales, we knew we had to keep feeding books into the channel.

But that's where things got difficult again. It just didn't seem possible to get enough books, quickly enough. But after everything we'd done to this point, I wasn't going to be stopped here. And I went into full "hands-on CEO" mode, trying to see how to juggle logistics to make things work.

The paper mill was in Glens Falls, NY. Once the paper had been made, it had to be trucked 2752 km to the printer in Winnipeg, Canada. Then the finished "book blocks" had to go 2225 km to the bindery in Toronto (or maybe there was an alternative bindery in Portland, OR, 2400 km away). And finally the bound books had to come to our warehouse in Illinois, or go directly to book distribution centers.

My archives contain a diagram I made trying to see how to connect these things together, particularly in view of the impending Canada Day holiday on July 1:

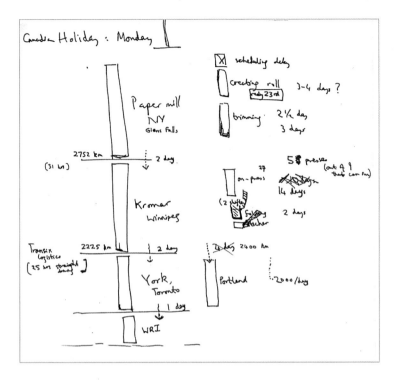

I have pages and pages of notes, with details of ink drying times (1 day), sheets of paper per skid (20,000), people needed per shift, and so on. But in the end we made it; with a lot of people's help, we got the books finished on time—and put on trucks, some of which were going to the distribution center for a major bookstore chain.

The trucks arrived. But then we heard nothing. Bookstores were reporting being out of stock. What was going on? At last it was figured out: multiple truckloads of books had somehow been misplaced at the distribution center. (How do you lose something that big?) And, yes, some sales momentum was lost. And so we didn't peak as high on bestseller lists as we might. Though hopefully in the end everyone who wanted an NKS book got one, no doubt oblivious to the logistical challenges involved in getting it to them.

## The Lost Epilog, and Other Outtakes from the Book

For more than a decade I basically poured everything I was doing into the NKS book. Well, at least that's the way I remember it. But going through my archives now, I realize I did quite a bit that never made it into the final NKS book. Particularly from the early years of the project, there are endless photographs—and investigations—of examples of complexity in nature, which never made it into Chapter 8. There are also lots of additional results about specific systems from the computational universe—as well as lots of details about history—that could have been notes to the notes, except I didn't have those.

Something I didn't remember is that in 1999—as the book was nearing completion—I considered adding a pictorial "Quick Summary" at the front of the book, here in draft form:

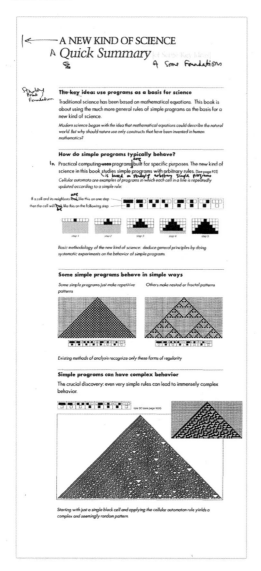
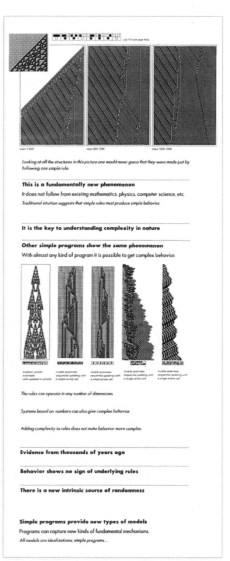

I'm not sure if this would have been a good idea, but in the end it effectively got replaced by the textual "An Outline of Basic Ideas" that appears at the very beginning of the book. Still, right when the book was being published, I did produce an "outside the book" pictorial 1-pager about Chapter 2 that saw quite a bit of use, especially for media briefings:

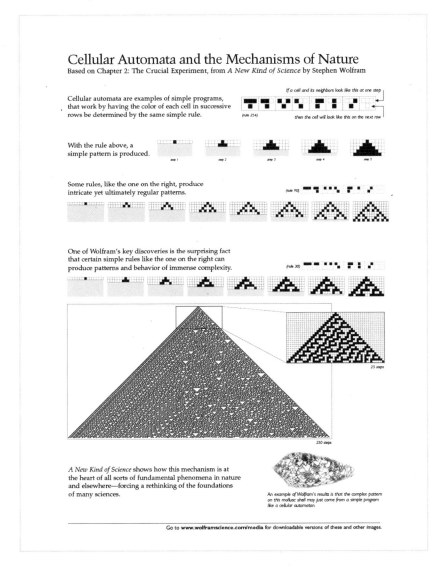

But as I was looking through my archives, my biggest "rediscovery" is the "Epilog" to the book. There are versions of it from quite early in the development of the book, but the last time it appears is in the December 15, 2000, draft—right before "Alpha 1". Then it's gone. Well, that is, until I just found it again:

So what's in this "lost epilog", with its intriguing title "The Future of the Science in This Book"? Different versions of it contain somewhat different fragmentary pieces of text. The version from late 1999, for example, begins:

> **Absorbing the Contents of the Book** [12.15]
>
> It has taken me nearly twenty years to develop the new kind of science that I describe in this book. And even though I have made every effort to present what I have done with the greatest of clarity I fully expect that many years will pass before it has all been widely absorbed.

Later it continues (the bracketed text gives alternative phrasings I was considering):

> **The Character of the New Intellectual Enterprise** [12.17]
>
> To create the new kind of science that I describe in this book over the past twenty years has been a largely solitary undertaking. But if what I have created is now to emerge as a major intellectual activity and a significant force in the intellectual world then inevitably a [whole community] large number of [other] people must become involved.
>
> Part of what this will achieve is to let a vast range of specific new discoveries and applications be made. But ultimately more important is that it will allow a [cultural structure] community of people to build up that can support [nurture] the kind of approach to science and scientific thinking that I have introduced in this book.
>
> What will make this successful is not so much its intellectual content, but the value system which defines what its participants consider important. [Is the point to solve the longest outstanding problem? To maximize commercial success?] Each of the existing fields of science has developed such a value system.

Some of what was in the "lost epilog" found its way into the Preface for the final book; some into a "General Note" entitled "Developing the new kind of science". But quite a lot never made it. It's often quite rough-hewn text—and almost just "notes to myself". But in a section entitled "What Should Be Done Now", there are, for example, suggestions like:

> What will survive best is the most general, abstract, yet simple. For this is one of the lessons of the history of science: mathematics has, for the most part, steadily built on its earlier results. But most other fields have gone through a series of revolutions in which older knowledge is discarded.

And there's a list of "principles" that aren't a bad summary of at least my general approach to research:

> Principles:
> - Always try to address the most obvious questions and find the simplest examples;
> - Try to understand the root causes of things; do not be satisfied with technical explanations;
> - Do not be bound by what has been done before, but try to understand it as fully as possible;
> - Explain what you have done as clearly as possible, and with as little infrastructure as possible

Later on there are some rough notes about what I thought might happen in the future:

> Phases of the new science (when they begin): [these are my expectations]
> - Absorption: try to understand what I have done in this book (first absorption completes in 2 years; more in 5 years)
> - Make the first round of extensions: (2 - 3 years; finished in 10-15 years)
> - Build major new directions (15 - 30 years)
> - Small early stage technological applications (4 - 10 years)
> - Major technological applications (10 - 25 years)
> - Become a part of everyday thought (4 - 10 years)
> - Become a standard part of basic science education (15 - 20 years)

It's a charming time-capsule-like item. But it's interesting to see how what I jotted down more than 20 years ago has actually panned out. And in fact I think much of it is surprisingly close to the mark. Plenty of small extensions did indeed get made in the first few years, with larger ones—both in studying abstract systems and in building practical models—coming later. (One notable extension was the 2,3 Turing machine universality proof at year 5, stimulated by our 2,3 Turing Machine Prize.)

How about "major new directions"? We're remarkably "on cue" there. At year 18 was our Physics Project, and from that has emerged the whole multicomputational paradigm, which I consider to be the next major direction building on the ideas of the

NKS book. I have to say that when I wrote down these expectations 20+ years ago, I didn't imagine that I would personally be involved in the "major new direction" I mentioned—but, unexpected as it has been, I feel very fortunate that that's the way it's worked out.

What about technology? Already at year 7 Wolfram|Alpha was in many ways a major "philosophical spinoff" of the NKS book. And although one doesn't know its detailed origins, the proof-of-work concept of bitcoin (which also first appeared at year 7) has fundamental connections to the idea of computational irreducibility. Meanwhile, the general methodology of searching the computational universe for useful programs is something that has continued to grow. And although the details are more complicated, the whole notion of deep learning in neural nets can also be thought of as related.

It's very hard to assess just what's happened in "becoming a part of everyday thought"—though it's been wonderful over the years to run into so many people who've told me how much the NKS book affected their way of thinking about things. But my impression is that—despite quite a few specific applications—the truly widespread absorption of ideas like computational irreducibility and their implications is a bit "behind schedule", though definitely now building well. (One piece of absorption that did happen in the 4–10 year window was into areas like art and architecture.)

What about education? 1D cellular automata have certainly become widely used as "do-a-little-extra" examples for both programming and math. But more serious integration of ideas from the NKS book as foundational elements of computational thinking—or as a kind of "pre-computer science"—is basically still a "work in progress".

Beyond the main text of the "lost epilog", I found something else: "Notes for the Epilog":

And after short (and unfinished) notes on "The sociology of the new science" and "The role of amateurs", there's the most significant "find": a list of altogether 283 "Open questions" for each of the chapters of the book, most still unanswered.

In preparation for our first Wolfram Summer School (then called the NKS Summer School) in June 2003, I worked on a more detailed version of something similar—but left it incomplete after getting up to the middle of Chapter 4, and didn't include much if anything from the "Notes to the Epilog" even though I'd been accumulating those for much of the time I worked on the book:

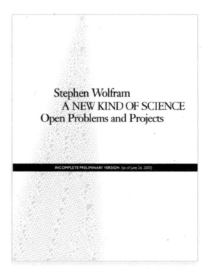

During the decade I worked on the NKS book I generated a vast amount of material. Most of it I kept in my still-very-much-extant computer filesystem, and while I can't say that I've reexamined everything there, my impression is that—perhaps apart from some "notes to the notes" material—a large fraction of what should have made it into the NKS book did. But in the course of working on the book there was definitely quite a bit of more ephemeral material. Some was preserved in my computer filesystem. But some was printed out and discarded, and some was simply handwritten. But all these years I've kept archive boxes of that material.

Some of those boxes have now been sealed for nearly 30 years. But I thought it'd be interesting to see what they contain. So I pulled out a box labeled 6/93–10/93. It's slightly the worse for wear after all these years, but what's inside is well preserved. I turn over a few pages of notes, printouts and ancient company memos (some sent as faxes). And then: what's this?

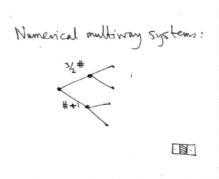

It's a note about multiway systems: things that are now central to the multicomputational paradigm I've just been pursuing. There's a brief comment about numerical multiway systems in the NKS book—but just last year, I wrote a whole 85-page "treatise" about them.

I turn over a few more pages. It feels a bit like a time warp. I just wrote about multiway Turing machines last year, and my very recent work on metamathematics is full of multiway string rewrites and their correspondence to mathematical proofs!

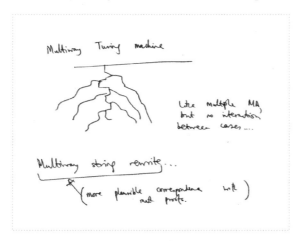

A few more pages and I get to:

> **A SCIENCE OF COMPLEXITY** | PART 2 DEVELOPING THE NEW INTUITION
>
> we now know that the same kind of behavior can also occur in continuous systems.
>
> **A New Intuition about Numbers**
>
> In this chapter we have seen that the traditional view of numbers can be quite misleading. So why has this view been so universally adopted in the past? The basic reason is that it fits in with methods such as calculus which have for many centuries dominated both pure mathematics and the applications of mathematics.
>
> Thousands of years ago, the first uses of numbers were probably for commerce, and only whole numbers and sometimes rational numbers were typically needed. But with the development of algebra and calculus over the course of the past five hundred years, it became increasingly convenient to assume that numbers could have a continuous range of possible sizes. And indeed, the notion of continuity became quite central to mathematics, and was adopted in essentially all the applications of mathematics.
>
> Yet even at the end of the nineteenth century, Cantor and others showed that completely discontinuous functions could in principle be constructed. Such functions were however for a long time viewed purely as mathematical curiosities, of no possible relevance to the applications of mathematics. But in fact it was from the study of these functions that both chaos theory and fractal geometry eventually emerged.
>
> In a sense, one of the achievements of algebra and calculus was to make it easy to find results about abstract sets of possible numbers. But in practical applications of mathematics, one in the end usually has to deal with individual numbers. And whether one does calculations by hand, by mechanical calculator, or by electronic computer, the abstract notion of a continuum of numbers is not particularly useful: one needs instead an explicit representation of numbers, typically in terms of digit sequences. (In the 1930's to 1960's some work was done on so-called analog computers which used physical processes to emulate continuous
>
> 56

It's not something that made it into the NKS book in that form—but last year I wrote a piece entitled "How Inevitable Is the Concept of Numbers?" which explores (in an admittedly modernized way) some of the exact same issues.

The Making of *A New Kind of Science*

A few more pages later I get to "timeless" graphics like these:

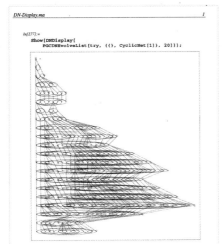 

But soon there's a charming reminder of the times:

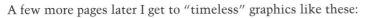

I've only gone through perhaps an inch of paper so far. And I'm getting to pages like these:

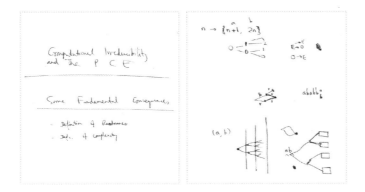

Yes, I'm still today investigating consequences of "computational irreducibility and the PCE (Principle of Computational Equivalence)". And just last year I used $n \mapsto \{n+1, 2n\}$ as a central example in writing about numerical multiway systems!

I've gone through perhaps 10% of one box—and there are more than 40 boxes in all. And I can't help but wonder what gems there may be in all these "outtakes" from the NKS book. But I'm also thankful that back when I was working on the NKS book I didn't try to pursue them all—or the decade I spent on the book might have stretched into more than a lifetime.

### And Now It's Out...

On May 14, 2002, the NKS book was finally published. In some ways the actual day of publication was quite anticlimactic. In modern times there'd be that moment of "making things live" (as there was, for example, for Wolfram|Alpha in 2009). But back then there'd been a big rush to get books to bookstores, but on the actual "day of publication" there wasn't much for me to do.

It had been a long journey getting to this point, though, and for example the acknowledgements at the front of the book listed 376 people who'd helped in one way or another over the decade devoted to writing the book, or in the years beforehand. But in terms of the physical production of the book one clue about what had been involved could be found on the very last page—its "Colophon":

> **Colophon**
>
> The original source for this book was created in FrameMaker, processed using an automated build system based on *Mathematica*, and output as PDF. (See also page 852.) The diagrams in the book were created using *Mathematica*, and the text for programs was automatically formatted by *Mathematica*, with both being imported as Encapsulated PostScript. Photographs were enhanced and processed using Photoshop and *Mathematica*. Index manipulation was done using *Mathematica* and IXgen.
>
> The fonts in the book are Trump Mediaeval, Palatino, Univers 45 and Gill Sans, with additional elements in Mathematica, Mathematica-Sans, Optima, Meridien and Janson.
>
> The book was printed on 50-pound Finch VHF paper on a sheet-fed press. It was imaged directly to plates at 2400 dpi, with halftones rendered using a 175-line screen with round dots angled at 45°. The binding was Smythe sewn.
>
> Book data: 1280 pages; 583,313 words (main text: 227,580, notes: 283,751); 2,799,438 characters; 973 illustrations; 1350 notes; 796 *Mathematica* programs; 14,967 index entries.
>
> Book designer: André Kuzniarek
>
> Cover designer: John Bonadies
>
> Additional designers: Jeremy Davis, Jody Jasinski
>
> Layout assistants: Larry Adelston, Richard Miske
>
> Production software developers: Andrew de Laix, Scott Koranda, Patrick Rice, Øyvind Tafjord
>
> Primary graphics finisher: Malgorzata Strzebonska
>
> Additional graphics finishers: Cookie Apichairuk, Hormozd Gahvari, Jay Hawkins, Nadya Markin, Jay Warendorff
>
> Custom font designers: Andy Hunt, André Kuzniarek
>
> Proofreaders: Jan Progen, Caroline Small and other members of the Wolfram Research Document Quality Assurance group
>
> Book program testers: Daniel Cranston, Bill Landis, Sung-il Pae, Niels Sondergaard
>
> Manufacturing manager: Brenda Skelly
>
> For other credits see pages xii–xiv.
>
> All diagrams and most photographs are original to this book. (Original photographs by the author and Chris Brown, Ian Collier, Matthew Cook, Andrew de Laix, Theodore Gray, André Kuzniarek, Conrad Wolfram.) Additional photographs are licensed from a variety of sources, believed to be in the public domain, or are courtesy of the following: Bildarchiv Preußischer Kulturbesitz; The Board of Trinity College Dublin; The British Museum; National Archaeological Museum, Greece (page 43); A. C. Charters; Yves Couder; P.E. Dimotakis; Peter Freymuth; H. Honji/S. Taneda; E. L. Koschmieder; Fred Landis; MIT Press; Maarten Rutgers Research Group; The Ohio State University; D. Howell Peregrine; Ephraim M. Sparrow (page 317); Ralph Buchsbaum; Jeremy Burgess/Science Photo Library. Photo Researchers; Chip Clark/The Smithsonian Institution; Christian de Duve; Eric Grave/Science Photo Library; Manfred Kage/Peter Arnold, Inc.; Lynn Margulis (page 385); John W. Forsythe/UTMB (page 416); Argos Group (page 993). Calligraphy on page 874 by Mamoun Sakkal.
>
> For legal notices see the copyright page.
>
> Printed by Kromar Printing Ltd, Winnipeg, Canada.

And, yes, as I've explained here, there was quite a story behind the simple paragraph: "The book was printed on 50-pound Finch VHF paper on a sheet-fed press. It was imaged directly to plates at 2400 dpi, with halftones rendered using a 175-line screen with round dots angled at 45°. The binding was Smythe sewn." And whatever other awards the book would win, it was rather lovely to win one for its creative use of paper:

So much about the NKS book was unusual. It was a book about new discoveries on the frontiers of science written for anyone to read. It was a book full of algorithmic pictures like none seen before. It was a book about science produced to a level of quality probably never equaled except by books about art. And it was a book that was published in a direct, entrepreneurial way without the intermediation of a standard large publishing company.

*Publishers Weekly* ran an interesting—and charmingly titled—piece purely about the "publishing dynamics" of the book:

# A New Kind of Self-Publishing

**by Charlotte Abbott**

Stephen Wolfram's *A New Kind of Science* sets a new standard in more ways than one

**book news**

It's a jaw-dropping accomplishment for an author to ship 50,000 copies of a self-published book to retailers on its publication date, let alone to hit #1 on Amazon.com in the first week on sale. But that's exactly what Stephen Wolfram did—and that was before *Time*, *Newsweek* and the *New York Times Book Review* covered the book.

Now, six weeks later, Wolfram's *A New Kind of Science* has hit even greater heights. Supported by brisk sales at online, chain and independent bookstores, it debuted at #16 on yesterday's *New York Times* extended list, following previous appearances on the *Wall St. Journal* and Barnes & Noble bestseller lists. It has also remained among Amazon's top 15 bestsellers since its May 12 publication, even though it has been sold without a discount through Amazon's Advantage program and has been shipping in two to three weeks. "Usually a book will spike, then drop a bit if it goes out of stock, but people just seem hellbent on getting this book," said Amazon bestsellers editor Tim Appelo.

Despite a retail price of $44.96 for the 1,197-page tome, which weighs in at 4½ pounds, wholesalers across the board are logging high demand as they wait for more stock to become available: 100,000 copies will be in print by the end of June. "It's a complicated book to produce, and now the issue is getting press time," explained publicist John Ekizian. The books are shipping directly from Kromar Printing in Winnipeg as they become available. Meanwhile, more media coverage is coming in the *Los Angeles Times*, *Washington Post*, *Boston Globe* and the *New York Times Magazine*. NPR's *Science Friday* will cover the book July 7, with national TV bookings likely to follow.

Having bypassed the peer review process and publication of extracts in scholarly journals that typically precede a major science book, *A New Kind of Science* has become Topic A on Internet chat sites for scientists, while drawing mixed reactions from critics. Most reviewers acknowledge the importance of Wolfram's thesis—that simple computer codes account more convincingly for complex natural phenomena than Newtonian mathematical equations. But several prominent reviews have disputed his claim to being the sole discoverer of this "new kind of science" and its validity when applied in diverse disciplines. While many reviewers have marveled at the eerily realistic images of snowflakes, tree branches and the pigmentation patterns on leopards that Wolfram has generated from simple algorithms and reproduced in the book, others maintain that they prove very little. Meanwhile, readers are entering the debate in droves: more than 100 postings on Amazon.com alternately praise and pillory the book in unusually erudite terms.

Heavy lifting: A 4.5 lb. bestseller takes off with more media to come.

For booksellers, the key questions are these: How quickly can they get stock and how many copies should they order? In answering the latter question, retailers must decide how the book's readability (or lack of it) will affect the scope of its appeal. Some, such as Karen Pennington at Kepler's Books and Magazines, dispute the claim that "any motivated reader should be able to plow through at least a few hundred pages before the details become too burdensome," as George Johnson declared in the *New York Times Book Review*. But having received 57 special orders for the book to date, she can't deny its sales potential among the highly educated, technically oriented readers who frequent her store near Stanford University. She compares the book to *Gödel, Escher, Bach* by Douglas T. Hofstadter, which became a bestseller in 1979 despite its intellectually challenging content. "It would be easy to make an overly conservative decision too early," she said. "But in college communities, this could become one of those books that just goes on and on."

**Not Just Any Author**

No stranger to unusual feats, Wolfram published his first paper on particle physics in 1975, when he was just 15 years old. The British-born prodigy earned his doctorate at CalTech at 20, becoming the youngest person ever to receive a MacArthur Foundation Fellowship the following year. By the time he was 26, he had innovated new ways to analyze such confounding physical phenomena as the

**WILEY IS MOVING!**

John Wiley & Sons, Inc., is moving their headquarters from

605 Third Avenue

in NYC to:

**111 River Street**

**Hoboken, NJ 07030**

**201-748-6000**

**www.wiley.com**

(Starting July 1st through August 5th)

WILEY

June 24, 2002 • PUBLISHERS WEEKLY 21

Used with permission of *Publishers Weekly*, from "A New Kind of Self-Publishing", Charlotte Abbott, 2002; permission conveyed through Copyright Clearance Center, Inc.

Just before the book was finally published, I'd signed some copies for friends, employees and people who'd contributed in one way or another to the book:

Shortly after the book was published, we decided to make a "commemorative poster", reproducing (small, but faithfully) every one of the pages that had taken so much effort to create:

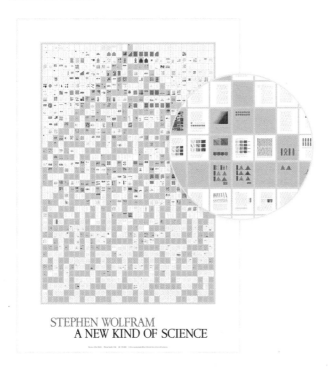

Then there were the "computational-irreducibility-inspired" bookmarks that I, for one, still use all the time:

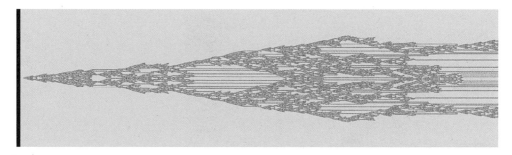

We carefully stored a virtual machine image of the environment used to produce the book (and, yes, that's how quite a few of the images here were made):

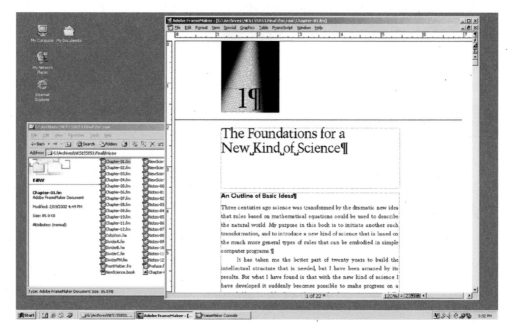

And over the years that followed we'd end up using the raw material for the book many times. Within a year there was "NKS Explorer"—a Wolfram Notebook system, distributed on CD-ROM, that served as a kind of virtual lab that let one (as it put it) "Experience the discoveries of *A New Kind of Science* on your own computer":

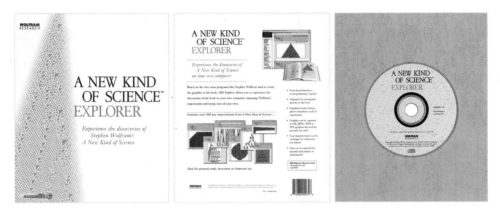

About five years later, more or less the same content would show up in the web-accessible Wolfram Demonstrations Project (and 10 years later, in its cloud version):

When the book came out, there was already a "wolframscience.com" website:

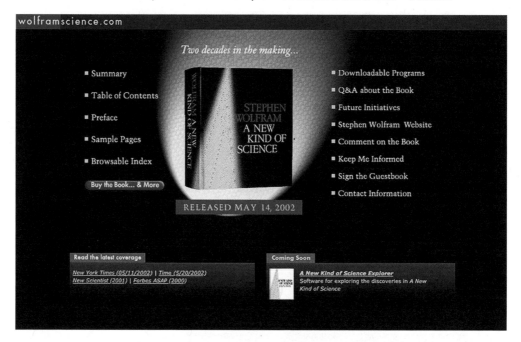

But in 2004 we were able to put a full version of the NKS book on the web:

In 2010 we made a version for the iPad:

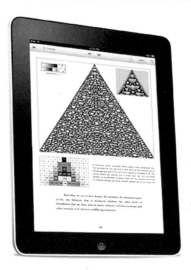

And in recent years there have followed all sorts of modernizations, especially on the web—with a bunch of new functionality just recently released:

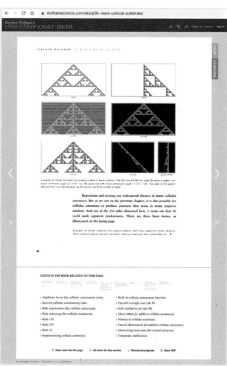

I went to great effort to write the NKS book to last, and I think it's fair to say—20 years out—that it very much has. The computational universe, of course, will be the same forever. And those pictures of the behavior of simple computational systems that occur throughout the book share the kind of fundamental timelessness that pictures of geometric constructions from antiquity do.

Of course, I knew that some things in the book would "date", most notably my references to technology—as I warned in one of the "General Notes" at the back of the book (though actually, 20 years later, notwithstanding "electronic address books" from page 643, and MP3 on page 1080 being described as a "recent" format, surprisingly little has yet changed):

> ■ **Technology references.** In an effort to make the main text of this book as timeless as possible, I have generally avoided referring to everyday systems whose character or name I expect will change as technology advances. Inevitably, however, I do discuss computers, even though I fully expect that some of the terms and concepts I use in connection with them will end up seeming dated in a matter of a few decades.

What about mistakes? For 20 years we've meticulously tracked them. And I think it's fair to say that all the careful checking we did originally really paid off, because in all the text and pictures in the book remarkably few errors have been found. For example, here's the list of everything in Chapter 4, indicating a few errors that were fixed in early printings—and a couple that remain, and that we are now fixing online:

### Chapter-04

| RT # | Location/Search String | Suggestion | Change | Date Implemented | Version Implemented | Comments |
|---|---|---|---|---|---|---|
| 2522 | p. 130, graphic (d) | Change f[n-2n] to f[n-2]. | f[n-2] | 5-14-02 | 1st edition/2nd printing | |
| 2728 | p. 130 | Remove the extra ] in the formula in the 4th graphic on this page. | | | | |
| 2637 | p. 140 | Change "If [r>s, ...]" to "[r>=s+1,...]". | "If [r>=s+1,...]" | 5-14-02 | 1st edition/2nd printing | |
| 2708 | p. 145-146 | Mistakenly identified graphic | | | | |
| 2578 | p. 148 | Change "is" to "it". | "it so far has been" | 2-07-02 | 1st edition/1st printing | |
| 2592 | p. 165 | Move "is" in front of "purely". | "is purely" | 5-14-02 | 1st edition/2nd printing | |

People ask me if there'll be a second edition of the NKS book. I say no. Yes, there are gradually starting to be more things one can say—and in the past couple of years the Wolfram Physics Project and the whole multicomputational paradigm has added significantly more. But there's nothing wrong with what's in the NKS book. It remains as valid and coherent as it was 20 years ago. And any "second-edition surgery" would run the risk of degrading its crispness and integrity—and detract from its unique perspective of presenting science at the time of its discovery.

But, OK, so all those NKS books that were printed on all those tons of paper from hemlock trees 20 years ago: what happened to them? Looking on the web today, one can find a few out there in the wild, sitting on bookshelves alongside a remarkable variety of other books:

I myself have many NKS books on my shelves (though admittedly a few more as convenient 2.5-inch "filler bookends"). And—at least when I'm in a "science phase"—I find myself using the online NKS book (if not a physical book) all the time, to see an example of some remarkable phenomenon in the computational universe, or to remind myself of some elaborate explanation or result that I put so much effort into finding all those years ago.

I consider the NKS book one the great achievements of my life—as well as one of the great "stepping-stone" points in my life, that was made possible by what I'd done before, and that in turn has made possible what I've done since. Twenty years later it's interesting to think back—as I've done here—on just what it took to produce the NKS book, and how all those individual steps that I worked so hard on for a decade came together to make the whole that is the NKS book.

To me it's a satisfying and inspiring story of what can be achieved with clear vision, sustained effort and a willingness to go where discoveries lead. And as I reflect on achievements of the past it makes me all the more enthusiastic about what's now possible—and why it's worth putting great effort today into what we can now build for the future.

# Charting a Course for "Complexity": Metamodeling, Ruliology and More

*Published September 23, 2021*

### "There's a Whole New Field to Build..."

For me the story began nearly 50 years ago—with what I saw as a great and fundamental mystery of science. We see all sorts of complexity in nature and elsewhere. But where does it come from? How is it made? There are so many examples. Snowflakes. Galaxies. Lifeforms. Turbulence. Did they all work differently? Or was there some common underlying cause? Some essential "phenomenon of complexity"?

It was 1980 when I began to seriously work on these questions. And at first I did so in the main scientific paradigm I knew: models based on mathematics and mathematical equations. I studied the approaches people had tried to use. Nonequilibrium thermodynamics. Synergetics. Nonlinear dynamics. Cybernetics. General systems theory. I imagined that the key question was: "Starting from disorder and randomness, how could spontaneous self-organization occur, to produce the complexity we see?" For somehow I assumed that complexity must be created as a kind of filtering of ubiquitous thermodynamic-like randomness in the world.

At first I didn't get very far. I could write down equations and do math. But there wasn't any real complexity in sight. But in a quirk of history that I now realize had tremendous significance, I had just spent a couple of years creating a big computer system that was ultimately a direct forerunner of our modern Wolfram Language. So for me it was obvious: if I couldn't figure out things myself with math, I should use a computer.

And there was something else: the computer system I'd built was a language that I'd realized (in a nod to my experience with reductionist physical science) would be the most powerful if it could be based on principles and primitives that were as minimal as possible. It had worked out very well for the language. And so when it came to complexity, it was natural to try to do the same thing. And to try to find the most minimal, most "meta" kind of model to use.

I didn't know just what magic ingredient I'd need in order to get complexity. But I thought I might as well start absolutely as simple as possible. And so it was that I set about running programs that I later learned were a simplified version of what had been called "cellular automata" before. I don't think it was even an hour before I realized that something very interesting was going on. I'd start from randomness, and "spontaneously" the programs would generate all sorts of complex patterns.

At first, it was experimental work. I'd make observations, cataloging and classifying what I saw. But soon I brought in analysis tools—from statistical mechanics, dynamical systems theory, statistics, wherever. And I figured out all sorts of things. But at the center of everything, there was still a crucial question: what was the essence of what I was seeing? And how did it connect to existing science?

I wanted to simplify still further. What if I didn't start from randomness, but instead started from the simplest possible "seed"? There were immediately patterns like fractals. But somehow I just assumed that a simple program, with simple rules, starting from a simple seed just didn't have what it took to make "true complexity". I had printouts (yes, that was still how it worked back then) that showed this wasn't true. But for a couple of years I somehow ignored them.

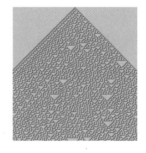

Then in 1984 I made my first high-resolution picture of rule 30. And I now couldn't get away from it: a simple rule and simple seed were making something that seemed extremely complex. But was it really that complex? Or was there some magic method of analysis that would immediately "crack" it? For months I looked for one. From mathematics. Mathematical physics. Computation theory. Cryptography. But I found nothing.

And slowly it began to dawn on me that I'd been fundamentally wrong in my basic intuition. And that in the world of simple programs—or at least cellular automata—complexity was actually easy to make. Could it really be that this was the secret that nature had been using all along to make complexity? I began to think it was at least a big part of it. I started to make connections to specific examples in crystal growth, fluid flow, biological forms and other places. But I also wanted to understand the fundamental principles of what was going on.

Simple programs could produce complex behavior. But why? It wasn't long before I realized something fundamental: that this was at its core a computational phenomenon. It wasn't something one could readily see with math. It required a different way of thinking about things. A fundamentally computational way.

At first I had imagined that having a program as a model of something was essentially just a convenience. But I realized that it wasn't. I realized that computational models were something fundamentally new, with their own conceptual framework, character and intuition. And as an example of that, I realized that they showed a new central phenomenon that I called computational irreducibility.

For several centuries, the tradition and aspiration of exact science had been to predict numbers that would say what a system would do. But what I realized is that in most of the computational universe of simple programs, you can't do that. Even if you know the rules for a system, you may still have to do an irreducible amount of computational work to figure out what it will do. And that's why its behavior will seem complex.

By 1985 I knew these things. And I was tremendously excited about their implications. I had got to this point by trying to solve the "problem of complexity". And it seemed only natural to label what could now be done as "complex systems theory": a theory of systems that show complexity, even from simple rules.

And so it was that in 1985 I began to promote the idea of a new field of "complex systems research", or, for short "complexity"—fueled by the discoveries I'd made about things like cellular automata.

Now that I know more about history I realize that the thrust of what I wanted to do had definite precursors, especially from the 1950s. For that was a time when the concepts of computing were first being worked out—and through approaches like cybernetics and the nascent area of artificial intelligence, people started exploring the broader scientific implications of computational ideas. But with no inkling of the phenomena I discovered decades later, this didn't seem terribly promising, and the effort was largely abandoned.

By the late 1970s, though, there were other initiatives emerging, particularly coming from mathematics and mathematical physics. Among them were fractals, catastrophe theory and chaos theory. Each in a different way explored some form of complexity. But all of them in a sense operated largely in the "comfort" of traditional mathematical ideas. And while they used computers as practical tools, they never made the jump to seeing computation as a core paradigm for thinking about science.

So what became of the "complex systems research" I championed in 1985? It's been 36 years now. Careers have come and gone. Several academic generations have passed by. Some things have developed well. Some things have not developed so well.

But I, for one, know much more than I did then. For me, my work in the early 1980s was a foundation for the whole tower of science and technology that I've spent my life since then building, most recently culminating in our Wolfram Physics Project and what in just the past few weeks I've called the multicomputational paradigm.

Nothing I've learned in these 36 years has dulled the strength and beauty of rule 30 and those early discoveries about complexity. But now I have so much more context, and a so-much-bigger conceptual framework—from which it's possible to see so much more about complexity and about its place and potential in science.

Back in 1985 I was pretty much a lone voice expressing the potential for studying complexity in science. Now there are perhaps a thousand scientific institutes around the world nominally focused on complexity. And my goal here is to share what I've learned and figured out about what's now possible to do under the banner of complexity.

There are exciting—and surprising—things. Some I was already beginning to think about in the 1980s. But others have only come into focus—or even become conceivable—as a result of very recent progress around our Physics Project and the formalism it has developed.

## The Emergence of a New Kind of Science

Back in 1985 I was tremendously excited about the potential for developing the field of complex systems research. It seemed as if there was a vast new domain that had suddenly been made accessible to scientific exploration. And in it I could see so much great science that could be done, and so many wonderful opportunities for so many people.

I myself was still only 25 years old. But I'd had some organizational experience, both leading a research group, and starting my first company. And I set about applying what I knew to complex systems research. By the following year, I'd founded the first research center and the first journal in the field (*Complex Systems*, still going strong after 35 years). (And I'd also done things like suggesting "complexity" as the theme for what became the Santa Fe Institute.) But somehow everything moved very slowly.

Despite my efforts, complex systems research wasn't a thing yet. It wasn't something universities were teaching; it wasn't something that was a category for funding. There were some applications for the field emerging. And there was tremendous pressure—particularly in the context of those applications—to shoehorn it into some existing area. Yes, it might have to take on the methodology of its "host" area. But at least it would have a home. But it really wasn't physics, or computer science, or math, or biology, or economics, or any known field. At least as I envisioned it, it was its own thing, with its own, new, emerging methodology. And that was what I really thought should be developed.

I was impatient to have it happen. And by late 1986 I'd decided the best path was just to try to do it myself—and to set up the best tools and the best environment for that. The result was Mathematica (and now the Wolfram Language), as well as Wolfram Research. For a few years the task of creating these entirely consumed me. But in 1991 I returned to basic science and set about continuing where I had left off five years earlier.

It was an exciting time. I quickly found that the phenomena I had discovered in cellular automata were quite general. I explored all sorts of different kinds of rules and programs, always trying to understand the essence of what they were doing. But every time, the core phenomena I found were the same. Computational irreducibility—as unexpected as it had been when I first saw it in cellular automata—was everywhere. And I soon realized that beneath what I was seeing, there was a deep and general principle—that I called the Principle of Computational Equivalence—that I now consider to be the most fundamental thing we know about the computational universe.

But what did these discoveries about simple programs and the computational universe apply to? My initial target had been immediately observable phenomena in the natural world. And I had somehow assumed that ideas like evolutionary adaptation or mathematical proof would be outside the domain. But as the years went by, I realized that the force of the Principle of Computational Equivalence was much greater than I'd ever imagined, and that it encompassed these things too.

I spent the 1990s exploring the computational universe and its applications, and steadily writing a book about what I was discovering. At first, in recognition of my original objective, I called the book *A Science of Complexity*. But by the mid-1990s I had realized that what I was doing far transcended the specific goal of understanding the phenomenon of complexity.

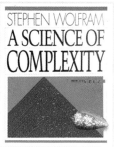

Instead, the core of what I was doing was to introduce a whole new kind of science, based on a new paradigm—essentially what I would now call the paradigm of computation. For three centuries, theoretical science had been dominated by the idea of using mathematical equations to describe the world. But now there was a new idea. The idea not of solving equations, but instead of setting up computational rules that could be explicitly run to represent and reproduce things in the world.

For three centuries theoretical models had been based on the fairly narrow set of constructs provided by mathematical equations, and particularly calculus. But now the whole computational universe of possible programs and possible rules was opened up as a source of raw material for making models.

But with this new power came a sobering realization. Out in the unrestricted computational universe, computational irreducibility is everywhere. So, yes, there was now a way to create models for many things. But to figure out the consequences of those models might take irreducible computational work.

Without the computational paradigm, systems that showed significant complexity had seemed quite inaccessible to science. But now there was an underlying way to model them, and to successfully reproduce the complexity of their behavior. But computational irreducibility was all over them, fundamentally limiting what could be predicted or understood about how they behave.

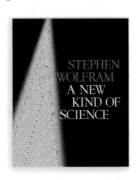

For more than a decade, I worked through the implications of these ideas, continually surprised at how many foundational questions across all sorts of fields they seemed to address. And particularly given the tools and technology I'd developed, I think I became pretty efficient at the research I did. And finally in 2002 I decided I'd pretty much "picked all the low-hanging fruit", and it was time to publish my magnum opus, titled—after what I considered to be its main intellectual thrust—*A New Kind of Science*.

The book was a mixture of pure, basic science about simple programs and what they do, together with a discussion of principles deduced from studying these programs, as well as applications to specific fields. If the original question had been "Where does complexity come from?" I felt I'd basically nailed that—and the book was now an exploration of what one could do in a science where the emergence of complexity from simplicity was just a feature of the deeper idea of introducing the computational paradigm as a foundation for a new kind of science.

I put tremendous effort into making the exposition in the book (in both words and pictures) as clear as possible—and contextualizing it with extensive historical research. And in general all this effort paid off excellently, allowing the message of the book to reach a very wide audience.

What did people take away from the book? Some were confused by its new paradigm ("Where are all the equations?"). Some saw it as a somewhat mysterious wellspring of new forms and structures ("Those are great pictures!"). But what many people saw in it was a thousand pages of evidence that simple programs—and computational rules—could be a rich and successful source of models and ideas for science.

It's hard to trace the exact chains of influence. But in the past two decades there's been a remarkable—if somewhat silent—transformation. For three hundred years, serious models in science had essentially always been based on mathematical equations. But in the short space of just twenty years that's all changed—and now the vast majority of new models are based not on equations but on programs. It's a dramatic and important paradigmatic change, whose implications are just beginning to be felt.

But what of complexity? In the past it was always a challenge to "get complexity" out of a model. Now—with computational models—it tends to be very easy. Complexity has gone from something mysterious and out of reach to something ubiquitous and commonplace. But what has that meant for the "study of complexity"? Well, that's a story with quite some complexity to it....

## The Growth of "Complexity"

From about 1984 to 1986 I put great effort into presenting and promoting the idea of "complex systems research". But by the time I basically left the field in 1986 to concentrate on technology for a few years, I hadn't seen much traction for the idea. A decade later, however, the story was quite different. I myself was quietly working away on what became *A New Kind of Science*. But elsewhere it seemed like my "marketing message" for complexity had firmly taken root, and there were complexity institutes starting to pop up all over the place.

What did people even mean by "complexity"? It often seemed to mean different things to different people. Sometimes it just meant "stuff in our field we haven't figured out yet". More often it meant "stuff that seems fundamental but we haven't been able to figure out". Pretty often there was some visualization component: "Look at how complex this plot seems!" But whatever exactly it might mean to different people, "complexity" was unquestionably becoming a popular science "brand", and there were plenty of people eager to affiliate with it—at the very least to give work they'd been doing for years a new air of modernity.

And while it was easy to be cynical about some of this, it had one very important positive consequence: "complexity" became a kind of banner for interdisciplinary work. As science had gotten bigger and more institutionalized, it inevitably became more siloed, with people in different departments at the same university routinely never having even met. But now people from all kinds of fields could say, "Yes, we run into complexity in our field", and with complexity as a banner they now had a reason to connect, and maybe even to form an institute together.

So what actually got done? Some of it I might summarize as "Yes, it's complex, but we can find something mathematical in it"—with a typical notion being the pursuit of some kind of power law formula. But the more important strand has been one that starts to actually take the computational paradigm on board—with the thrust typically being "We can write a program to reproduce what we're looking at".

And one of the great feelings of power has been that even in fields—like the social sciences—where there haven't really been much more than "verbal" models before, it's now appeared possible to get models that at least seem much more "scientific". Sometimes the models have been purely empirical ("Look, there's a power law!"). Sometimes they have been based on constructing programs to reproduce behavior.

The definition of success has often been a bit questionable, however. Yes, there's a program that shows some features of whatever system one's looking at. But how complicated is the program? How much of what's coming out is basically just being put right into the program? For mathematical models, people have long had familiarity with questions like "How many parameters does that model have?". But when it comes to programs, there's been a tendency just to put more and more into them without doing much accounting of it.

And then there's the matter of complexity. Let's say whatever one's trying to model shows complexity. Then often the thinking seems to be that to get that complexity out, there's a need to somehow have enough complexity in the model. And when complexity does manage to come out, there's a feeling that this is some kind of triumph, and evidence that the model is on the right track.

But actually—as I discovered in studying the computational universe of simple programs—this really isn't the right intuition at all. Because it fundamentally doesn't take into account computational irreducibility. And knowing that computational irreducibility is ubiquitous, we know that complexity is too. It's not something special and "on the right track" that's making a model produce complexity; instead producing complexity is just something a very wide range of computational models naturally do.

Still, the general field and brand of complexity continued to gain traction. Back in 1986 my *Complex Systems* had been the only journal devoted to complex systems research. By the late 2010s there were dozens of journals in the field. And my original efforts from the 1980s to promote the study of complexity had been thoroughly dwarfed by a whole "complexity industry" that had grown up. But looking at what's been done, I feel like there's something important that's missing. Yes, it's wonderful that there's been so much "complexity activity". But it feels scattered and incoherent—and without a strong common thread.

## Returning to the Foundations of Complexity

There's a vast amount that's now been done under the banner of complexity. But how does it fit together? And what are its intellectual underpinnings? The dynamics of academia has led most of the ongoing activity of complexity research to be about specific applications in specific fields—and not really to concern itself with what basic science might lie underneath, and what the "foundations of complexity" might be.

But the great power of basic science is the economy of scale it brings. Find one principle of basic science and it can inform a vast range of different specific applications, that would otherwise each have to be explored on their own. Learn one principle of basic science and you immediately know something that subsumes all sorts of particular things you would otherwise separately have to learn.

So what about complexity? Is there something underneath all those specifics that one can view as a coherent "basic science of complexity"—and for example the raw material for something like a course on the "Foundations of Complexity"? At first it might not be obvious where to look for this. But there's immediately a big clue. And it's what is in a sense the biggest "meta discovery" of the study of complexity over the past few decades: that across all kinds of systems, computational models work.

So then one's led to the question of what the basic science of computational models—or computational systems in general—might be. But that's precisely what my work on the computational universe of simple programs—and my book *A New Kind of Science*—are about. They're about the core basic science of the computational universe, and the principles it involves—in a sense the foundational science of computation.

It's important, by the way, to distinguish this from computer science. Computer science is about programs and computations that we humans construct for certain purposes. But the foundational science we need is instead about programs and computations "in the wild"—and about what's out there in general in the computational universe, independent of whether we humans would have a reason to construct or use it.

It's a very abstract kind of thing. That—like pure mathematics—can be studied completely on its own, without reference to any particular application. And in fact the analogy to pure mathematics is an apt one. Because just as pure mathematics is in a sense the abstract underpinning for the mathematical sciences and the whole mathematical paradigm for representing the world, so now our foundational science of computation is the abstract underpinning for the computational paradigm for representing the world—and for all the "computational X" fields that flow from it.

So, yes, there is a core basic science of complexity. And it's also essentially the foundational science of computation. And by studying this, we can bring together all sorts of seemingly disparate issues that arise in the study of complexity in different systems. Everywhere we'll see computational irreducibility. Everywhere we'll see intrinsic randomness generation. Everywhere we'll see the effects of the Principle of Computational Equivalence. These are general, abstract things from pure basic science. They're the intellectual underpinnings of the study of complexity—the "foundations of complexity".

## Metamodeling and the Metatheory of Models

> **metamodeling** (*n.*) the metascience of finding minimal models for models

I was at a complexity conference once, talking to someone who was modeling fish and their behavior. Proudly the person showed me his simulated fish tank. "How many parameters does this involve?", I asked. "About 90", he said. "My gosh", I said, "with that many parameters, you could put an elephant in your fish tank too!"

If one wanted to make a simulated fish tank display just for people to watch, then having all those parameters might be just fine. But it's not so helpful if one wants to understand the science of fish. The fish have different shapes. The fish swim around in different configurations. What are the core things that lead to what we see?

To answer that, we have to drill down: we have to find the essence of fish shape, or fish behavior.

At first, if confronted with complexity, we might say "It's hopeless, we'll never find the essence of what's going on—it's all too complicated". But the whole point is that we know that in the computational universe of possible programs, there can in fact be simple programs with simple rules that lead to immense complexity. So even though there's immense complexity in behavior we see, underneath it all there can still be something simple and understandable.

In a sense, the concept of taking phenomena and drilling down to find their underlying essential causes is at the heart of reductionist science. But as this has traditionally been practiced, it's relied on being able to see one's way through this "drilling down" process, or in effect, to explicitly do reverse engineering. But a big lesson of the computational paradigm is the phenomenon of computational irreducibility—and the "irreducible distance" that can exist between rules and the behavior they produce.

It's a double-edged thing, however. Yes, it's hard to drill down through computational irreducibility. But in the end the details of what's underneath may not matter so much; the main features one sees may just be generic reflections of the phenomenon of computational irreducibility.

Still, there are normally structural features of the underlying models (or their interpretations) that matter for particular applications. Is one dealing with something on a 2D grid? Are there nonlocal effects in the system? Is there directionality to the states of the system? And so on.

If one looks at the literature of complexity, one finds all sorts of models for all sorts of systems. And often—like the fish example—the models are very complicated. But the question is: are there simpler models lurking underneath? Models simple enough that one can readily understand at least their basic rules and structure. Models simple enough that it's plausible that they could be useful for other systems as well.

To find such things is in a sense an exercise in what one can call "metamodeling": trying to make a model of a model, doing reductionist science not on observations of the world, but on the structure of models.

When I first worked on the problem of complexity, one of the main things I did was a piece of metamodeling. I was looking at models for a whole variety of phenomena, from snowflake growth to self-gravitating gases to neural nets. But what I did was to try to identify an underlying "metamodel" that would cover all of them. And what I came up with was simple cellular automata (which, by the way, don't cover everything I had been looking at, but turn out to be very interesting anyway).

As I think about it now, I realize that the activity of metamodeling is not a common one in science. (In mathematics, one could argue that something like categorification is somewhat analogous.) But to me personally, metamodeling has seemed very natural—because it's very much like something I've done for a very long time, which is language design.

What's involved in language design? You start off from a whole collection of computations, and descriptions of how to do them. And then you try to drill down to identify a small set of primitives that let you conveniently build up those computations. Just like metamodeling is about removing all the "hairy" parts of models to get to their minimal, primitive forms, so also language design is about doing that for computations and computational structures.

In both cases there's a certain art to it. Because in both cases the consumers of those minimal forms are humans. And it's to humans that they need to seem "simple" and understandable. Some of the practical definition of simplicity has to do with history. What, for example, has become familiar, or are there words for? Some is more about human perception. What can be represented by a diagram that our visual processing system can readily absorb?

But once one's found something minimal, the great value of it is that it tends to be very general. Whereas a detailed "hairy" model tends to have all sorts of features specific to a particular system, a simple model tends to be applicable to all sorts of systems. So by doing the metamodeling, and finding the simplest "common" model, one is effectively deriving something that will have the greatest leverage.

I've seen this quite dramatically with cellular automata over the past forty years. Cellular automata are in a sense minimal models in which there's a definite (discrete) structure for space and time and a finite number of states associated with each discrete cell. And it's been remarkable how many different kinds of systems can successfully be modeled by cellular automata. So that for example of the 256 very simplest 2-color nearest-neighbor 1D rules, a significant fraction have found application somewhere, and many have found several (often completely different) applications.

I have to say that I haven't explicitly thought of myself as pursuing "metamodeling" in the past (and I only just invented the term!). But I believe it's an important technique and idea. And it's one that can "mine" the specific modeling achievements of work on complexity and bring them to a broader and more foundational level.

In *A New Kind of Science* I cataloged and studied minimal kinds of models of many types. And in the twenty years since *A New Kind of Science* was finished, I have only seen a modest number of new minimal models (though I haven't been looking for them with the focus that metamodeling now brings). But recently, I have another major example of what I now call metamodeling. For our Physics Project we've developed a particular class of models based on multiway hypergraph rewriting. But I've recently realized that there's metamodeling to do here, and the result has been the general concept of multicomputation and multicomputational models.

Returning to complexity, one can imagine taking all the academic papers in the field and identifying the models they use—and then trying to do metamodeling to classify and boil down these models. Often, I suspect, the resulting minimal classes of models will be ones we've already seen (and that, for example, appear in *A New Kind of Science*). But occasionally they will be new: in a sense new primitives for the language of modeling, and new "metascientific" output from the study of complexity.

## The Pure Basic Science of Ruliology

> **ruliology** (*n.*) the pure basic science of what simple rules do

If one sets up a system to follow a particular set of simple rules, what will the system do? Or, put another way, how do all those simple programs out there in the computational universe of possible programs behave?

These are pure, abstract questions of basic science. They're questions one's led to ask when one's operating in the computational paradigm that I describe in *A New Kind of Science*. But at some level they're questions about the specific science of what abstract rules (that we can describe as programs) do.

What is that science? It's not computer science, because that would be about programs we construct for particular purposes, rather than ones that are just "out there in the wilds of the computational universe". It's not (as such) mathematics, because it's all about "seeing what rules do" rather than finding frameworks in which things can be proved. And in the end, it's clear it's actually a new science—that's rich and broad, and that I, at least, have had the pleasure of practicing for forty years.

But what should this science be called? I've wondered about this for decades. I've filled so many pages with possible names. Could it be based on Greek or Latin words associated with rules? Those are *arch-* and *reg-*: very well-trafficked roots. What about words associated with computation? That'd be *logis-* or *calc-*. None of these seem to work. But—in something akin to the process of metamodeling—we can ask: What is the essence of what we want to communicate in the word?

It's all about studying rules, and what their consequences are. So why not the simple and obvious "ruliology"? Yes, it's a new and slightly unusual-sounding word. But I think it does well at communicating what this science that I've enjoyed for so long is about. And I, for one, will be pleased to call myself a "ruliologist".

But what is ruliology really about? It's a pure, basic science—and a very clean and precise one. It's about setting up abstract rules, and then seeing what they do. There's no "wiggle room". No issue with "reproducibility". You run a rule, and it does what it does. The same every time.

What does the rule 73 cellular automaton starting from a single black cell do? What does some particular Turing machine do? What about some particular multiway string substitution system? These are specific questions of ruliology.

At first you might just do the computation, and visualize the result. But maybe you notice some particular feature. And then you can use whatever methods it takes to get a specific ruliological result—and to establish, for example, that in the rule 73 pattern, black cells appear only in odd-length blocks.

Ruliology tends to start with specific cases of specific rules. But then it generalizes, looking at broader ranges of cases for a particular rule, or whole classes of rules. And it always has concrete things to do—visualizing behavior, measuring specific features, and so on.

But ruliology quickly comes face to face with computational irreducibility. What does some particular case of some particular rule eventually do? That may require an irreducible amount of computational effort to find out—and if one insists on knowing what amounts to a general truly infinite-time result, it may be formally undecidable. It's the same story with looking at different cases of a rule, or different rules. Is there any case that does this? Or any rule that does it?

What's remarkable to me—even after 40 years of ruliology—is how many surprises there end up being. You have some particular kind of rule. And it looks as if it's only going to behave in some particular way. But no, eventually you find a case where it does something completely different, and unexpected. And, yes, this is in effect computational irreducibility reaching into what one's seeing.

Sometimes I've thought of ruliology as being at first a bit like natural history. You're exploring the world of simple programs, finding what strange creatures exist in it—and capturing them for study. (And, yes, in actual biological natural history, the diversity of what one sees is presumably at its core exactly the same computational phenomenon we see in abstract ruliology.)

So how does ruliology relate to complexity? It's a core part—and in fact the most fundamental part—of studying the foundations of complexity. Ruliology is like studying complexity at its ultimate source. And about seeing just how complexity is generated from its simplest origins.

Ruliology is what builds raw material—and intuition—for making models. It's what shows us what's possible in the computational universe, and what we can use to model—and understand—the systems we study.

In metamodeling we're going from models that have been constructed, and drilling down to see what's underneath them. In ruliology we're in a sense going the other way, building up from the minimal foundations to see what can happen.

In some ways, ruliology is like natural science. It's taking the computational universe as an abstracted analog of nature, and studying how things work in it. But in other ways, ruliology is something more generative than natural science: because within the science itself, it's thinking not only about what is, but also about what can abstractly be generated.

Ruliology in some ways starts as an experimental science, and in some ways is abstract and theoretical from the beginning. It's experimental because it's often concerned with just running simple programs and seeing what they do (and in general, computational irreducibility suggests you often can't do better). But it's abstract and theoretical in the sense that what's being run is not some actual thing in the natural world, with all its details and approximations, but something completely precise, defined and computational.

Like natural science, ruliology starts from observations—but then builds up to theories and principles. Long ago I found a simple classification of cellular automata (starting from random initial conditions)—somehow reminiscent of identifying solids, liquids and gases, or different kingdoms of organisms. But beyond such classifications, there are also much broader principles—with the most important, I believe, being the Principle of Computational Equivalence.

The everyday course of doing ruliology doesn't require engaging directly with the whole Principle of Computational Equivalence. But throughout ruliology, the principle is crucial in guiding intuition, and having an idea of what to expect. And, by the way, it's from ruliology that we can get evidence (like the universality of rule 110, and of the 2,3 Turing machine) for the broad validity of the principle.

I've been doing ruliology (though not by that name) for forty years. And I've done a lot of it. In fact, it's probably been my top methodology in everything I've done in science. It's what led me to understand the origins of complexity, first in cellular automata. It's what led me to formulate the general ideas in *A New Kind of Science*. And it's what gave me the intuition and impetus to launch our new Physics Project.

I find ruliology deeply elegant, and satisfying. There's something very aesthetic—at least to me—about the purity of just seeing what simple rules do. (And it doesn't hurt that they often make very pleasing images.) It's also satisfying when one can go from so little and get so much—and do so automatically, just by running something on a computer.

And as well I like the fundamental permanence of ruliology. If one's dealing with the simplest rules of some type, they're going to be foundational not only now, but forever. It's like simple mathematical constructs—like the icosahedron. There were icosahedral dice in ancient Egypt. But when we find them today, their shapes still seem completely modern—because the icosahedron is something fundamental and timeless. Just like the rule 30 pattern or countless other discoveries in ruliology.

In a sense perhaps one of the biggest surprises is that ruliology is such a comparatively new activity. But as I cataloged in *A New Kind of Science*, it has precursors going back hundreds and perhaps thousands of years. But without the whole paradigm of *A New Kind of Science*, there wasn't a context to understand why ruliology is so significant.

So what constitutes a good piece of ruliology? I think it's all about simplicity and minimality. The best ruliology happens after metamodeling is finished—and one's really dealing with the simplest, most minimal class of rules of some particular type. In my efforts to do ruliology, for example in *A New Kind of Science*, I like to be able to "explain" the rules I'm using just by an explicit diagram, if possible with no words needed.

Then it's important to show what the rules do—as explicitly as possible. Sometimes—as in cellular automata—there's a very obvious visual representation that can be used.

But in other cases it's important to do the work to find some scheme for visualization that's as explicit as possible, and that both shows the whole of what's going on and doesn't introduce distracting or arbitrary additional elements.

It's amazing how often in doing ruliology I'll end up making an array of thumbnail images of how certain rules behave. And, again, the explicitness of this is important. Yes, one often wants to do various kinds of filtering, say of rules. But in the end I've found that one needs to just look at what happens. Because that's the only way to successfully notice the unexpected, and to get a sense of the irreducible complexity of what's out there in the computational universe of possible rules.

When I see papers that report what amounts to ruliology, I always like it when there are explicit pictures. I'm disappointed if all I see are formal definitions, or plots with curves on them. It's an inevitable consequence of computational irreducibility that in doing good ruliology, one has to look at things more explicitly.

One of the great things about ruliology as a field of study is how easy it is to explore new territory. The computational universe contains an infinite number of possible rules. And even among ones that one might consider "simple", there are inevitably astronomically many on any human scale. But, OK, if one explores some particular ruliological system, what of it?

It's a bit like chemistry where one explores properties of some particular molecule. Exploring some particular class of rules, you may be lucky enough to come upon some new phenomenon, or understand some new general principle. But what you know you'll be doing is systematically adding to the body of knowledge in ruliology.

Why is that important? For a start, ruliology is what provides the raw material for making models, so you're in effect creating a template for some potential future model. And in addition, when it comes to technology, an important approach that I've discussed (and used) quite extensively involves "mining" the computational universe for "technologically useful" programs. And good ruliology is crucial in helping to make that feasible.

It's a bit like creating technology in the physical universe. It was crucial, for example, that good physics and chemistry had been done on liquid crystals. Because that's what allowed them to be identified—and used—in making displays.

Beyond its "pragmatic" value for models and for technology, another thing ruliology does is to provide "empirical raw material" for making broader theories about the computational universe. When I discovered the Principle of Computational Equivalence, it was as a result of several years of detailed ruliology on particular types of rules. And good ruliology is what prepares and catalogs examples from which theoretical advances can be made.

It's worth mentioning that there's a certain tendency to want to "nail down ruliology" using, for example, mathematics. And sometimes it's possible to derive a nice summary of ruliological results using, say, some piece of discrete mathematics. But it's remarkable how quickly the mathematics tends to get out of hand, with even a very simple rule having behavior that can only be captured by large amounts of obscure mathematics. But of course that's in a sense just computational irreducibility rearing its head. And showing that mathematics is not the methodology to use—and that instead something new is needed. Which is precisely where ruliology comes in.

I've spent many years defining the character and subject matter of what I'm now calling ruliology. But there's something else I've done too, which is to build a large tower of practical technology for actually doing ruliology. It's taken more than forty years to build up to what's now the full-scale computational language that is the Wolfram Language. But all that time, I was using what we were building to do ruliology.

The Wolfram Language is great and important for many things. But when it comes to ruliology, it's simply a perfect fit. Of course it's got lots of relevant built-in features. Like visualization, graph manipulation, etc., as well as immediate support for systems like cellular automata, substitution systems and Turing machines. But what's even more important is that its fundamental symbolic structure gives it an explicit way to represent—and run—essentially any computational rule.

In doing practical ruliological explorations—and for example searching the computational universe—it's also useful to have immediate support for things like parallel computation. But another crucial aspect of the Wolfram Language for doing practical ruliology is the concept of notebooks and computable documents. Notebooks let one organize both the process of research and the presentation of its results.

I've been accumulating research notebooks about ruliology for more than 30 years now—with textual notes, images of behavior, and code. And it's a great thing. Because the stability of the Wolfram Language (and its notebook format) means that I can immediately go back to something I did 30 years ago, run the code, and build on it. And when it comes to presenting results, I can do it as a computational essay, created in a notebook—in which the task of exposition is shared between text, pictures and computational language code.

In a traditional technical paper based on the mathematical paradigm, the formal part of the presentation will normally use mathematical notation. But for ruliology (as for "computational X" fields) what one needs instead is computational notation, or rather computational language—which is exactly what the Wolfram Language provides. And in a good piece of ruliology—and ruliology presentation—the notation should be simple, clear and elegant. And because it's in computational language, it's not just something people read; it's also something that can immediately be executed or integrated somewhere else.

What should the future of ruliology be? It's a huge, wide-open field. In which there are many careers to be made, and immense numbers of papers and theses and books that can be written—that will build up a body of knowledge that advances not just the pure, basic science of the computational universe but also all the science and technology that flows from it.

## Philosophy and the Foundations of Complexity

How should the phenomenon of complexity affect one's worldview, and one's general way of thinking about things? It's a bit of a roller-coaster-like ride. When first confronted with complexity in a system, one might think "There doesn't seem to be any science to that". But then with great effort it may turn out to be possible to "drill down" and find the underlying rules for the system, and perhaps they'll even be quite simple. And at that point we might think "OK, science has got this one"—we've solved it.

But that ignores computational irreducibility. And computational irreducibility implies that even though we may know the underlying rules, that doesn't mean we can necessarily "scientifically predict" what the system will do; instead, it may take an irreducible amount of computational work to figure it out.

Yes, you may have a model that correctly captures the underlying rules for a system—and even explains the overall complexity in the behavior of the system. But that absolutely does not mean that you can successfully make specific predictions about what the system will do. Because computational irreducibility gets in the way, essentially "eating away the power of science from the inside"—as an inevitable formal fact about how systems based on the computational paradigm typically behave.

But in a sense even the very phenomenon of computational irreducibility—and even more so, the Principle of Computational Equivalence—give us ways to reason and think about things. It's a bit like in evolutionary biology, or in economics, where there are principles that don't specifically define predictions, but do give us ways to reason and think about things.

So what are some conceptual and philosophical consequences of computational irreducibility? One thing it does is to explain ubiquitous apparent randomness in the world, and say why it must happen—or at least must be perceived to happen by computationally bounded observers like us. And another thing it does is to tell us something about the perception of free will. Even if the underlying rules for a system (such as us humans) are deterministic, there can be an inevitable layer of computational irreducibility which makes the system still seem to a computationally bounded observer to be "free".

Metamodeling and ruliology are in effect the extensions of traditional science needed to handle the phenomenon of complexity. But what about extensions to philosophy?

For that one must think not just about the phenomenology of complexity, but really about its foundations. And that's where I think one inevitably runs into the whole computational paradigm, with all its intellectual implications. So, yes, there's a "philosophy of complexity", but it's really the "philosophy of the computational paradigm".

I started to explore this towards the end of *A New Kind of Science*. But there's much more to be done, and it's yet something else that can be reached by serious study of the foundations of complexity.

## Multicomputation and the (Surprise) Return of Reducibility

Computational irreducibility is a very strong phenomenon, that in a sense pervades the computational universe. But within computational irreducibility, there must always be pockets—or slices—of computational reducibility: aspects of a system that are amenable to a reduced description. And for example in doing ruliology, part of the effort is to catalog the computational reducibility one finds.

But in typical ruliology—or, for example, a random sampling of the computational universe of possible programs—computational reducibility is at best a scattered phenomenon. It's not something one can count on seeing. But there's something confusing about this when it comes to thinking about our universe, and our experience of it. Because perhaps the most striking fact about our universe—and indeed the one that leads to the possibility of what we normally call science—is that there's order in what happens in it.

Yet even if the universe ultimately operates at the lowest level according to simple rules, we might expect that at our level, all we would see is rampant computational irreducibility. But in our recent Physics Project there has been a big surprise. Because with the structure of the models we used, it seemed that within all that computational irreducibility, we were always seeing certain slices of reducibility—that turn out to correspond to the major known laws of physics: general relativity and quantum mechanics.

A more careful examination showed that what was picking out this computational reducibility was really the combination of two things. First, a certain general structure to the underlying model. And second, certain rather general features of us as observers of the system.

In the usual computational paradigm, one imagines rules that are successively applied to determine how the state of a system should evolve in time. But our Physics Project needed a new paradigm—that I've recently called the multicomputational paradigm—in which there can be many possible states of a system evolving in effect on many possible interwoven threads of time. In the computational paradigm, one can always identify the particular state reached after a certain amount of evolution. But in the multicomputational paradigm, it takes an observer to define how a "perceived state" should be extracted from all the possible threads of time.

In the multicomputational paradigm, the actual evolution on all the threads of time will show all sorts of computational irreducibility. But somehow what an observer like us perceives has "smoothed" all of that out. And what's left is something that's a reflection of the core structure of the underlying multicomputational rules. And that turns out to show a certain set of emergent "physics-like laws".

It's all an important piece of metamodeling. We started from a model intended to capture fundamental physics. But we've been able to "drill down" to find the essential "primitive structure" underneath—which turns out to be the idea of multicomputation. And wherever multicomputation occurs, we can expect that there will be computational reducibility and emergent physics-like laws, at least for certain kinds of observers.

So how does this relate to complexity? Well, when systems fundamentally follow the computational paradigm—with standard computational models—they'll tend to show computational irreducibility and complexity. But if instead they follow the multicomputational paradigm, then there'll be emergent laws to discover in them.

There are all sorts of fields—like economics, linguistics, molecular biology, immunology, etc.—where I have recently come to suspect that there may be good multicomputational models to be made. And in these fields, yes, there will be complexity to be seen. But the multicomputational paradigm suggests that there will also be definite regularities and emergent laws. So in a sense from "within complexity" there will inexorably emerge a certain simplicity. So that if one "observes the right things" one can potentially find what amount to "ordinary scientific laws".

It's a curious twist in the story of complexity, and one that I, for one, did not see coming. Back in the early 1980s when I was first working on complexity, I used to talk about finding "scientific laws of complexity". And at some level computational irreducibility and the Principle of Computational Equivalence are very general such laws—that were at first very surprising to see.

But what we've discovered is that in the multicomputational paradigm, there's another surprise: complexity can produce simplicity. But not just any simplicity. Simplicity that specifically follows physics-like laws. And that for a variety of fields might indeed give us something we could consider to be "scientific laws of complexity".

## What Should Happen Now

It's a wonderful thing to see something go from "just an idea" to a whole, developed ecosystem in the world. But that's what's happened over the past forty years with the concept of doing science around the phenomenon of complexity. And over that time countless "workflows" associated with particular applications have been developed—and there's been all sorts of activity in all sorts of areas. But now I think it's time to take stock of what's been achieved—and see what might be possible going forward.

I myself have not been much involved in the "daily work of the complexity field" since my early efforts in the 1980s. And perhaps that distance makes it easier to see what lies ahead. For, yes, by now there's plenty of understanding of how to apply "complexity-inspired methodology" (and computational models) in particular areas. But the great opportunity is to turbocharge all this by focusing again on the "foundations of complexity"—and bringing the basic science that arises from that to bear on all the various applications whose "workflows" have now been defined.

But what is that basic science? Its great "symptom" is complexity. But there's much more to it than that. It's heavily based on the computational paradigm. And it's full of deep and powerful ideas and methods. And I've been thinking about it for more than forty years. But it's only very recently—particularly based on what I've learned from our Physics Project—that I think I see with true clarity just how that science should be defined and pursued.

First, there's what I'm calling here metamodeling: going from specific models constructed for particular applications, and working out what the underlying more minimal and more general models are. And second, there's what I'm calling ruliology: the study of what possible rules (or possible programs) in the computational universe do.

Metamodeling is a kind of "meta" analog of science, probably most directly related to activities like computational language design. Ruliology is a pure, basic science, a bit like pure mathematics, but based on a very different methodology.

In both metamodeling and ruliology there is much of great value to do. And even after more than forty years of pursuing what I'm now calling ruliology, I feel as if I've only just scratched the surface of what's possible.

Applications under the banner of complexity will come and go as different fields and their objectives ebb and flow. But both metamodeling and ruliology have a certain purity, and clear anchoring to intellectual bedrock. And so we can expect that whatever is discovered there will—like the discoveries of pure mathematics—be part of the permanent corpus of theoretical knowledge.

Hovering over all of what we might study around complexity is the phenomenon of computational irreducibility. But within that irreducibility are pockets and slices of reducibility. And informed by our Physics Project, we now know that multicomputational systems can be expected to expose to observers like us what amount to physics-like laws—in effect leveraging the phenomenon of complexity to deliver accessible scientific laws.

Complexity is a field that fundamentally rests on the computational paradigm—and in a sense when we see complexity what is really happening is that some lump of irreducible computation is being exposed. So at its core, the study of complexity is a study of irreducible computation. It's computation whose details are irreducibly hard to figure out. But which we can reason about, and which, for example, we can also use for technology.

Even forty years ago, the fundamental origin of complexity still seemed like a complete mystery—a great secret of nature. But now through the computational paradigm, I think we have a clear notion of where complexity fundamentally comes from. And by leveraging the basic science of the computational universe—and what I'm now calling metamodeling and ruliology—there's a tremendous opportunity that now exists to dramatically advance everything that's been done under the banner of complexity.

The first phase of "complexity" is complete. The ecosystem is built. The applications are identified. The workflows are defined. And now it's time to return to the foundations of complexity. And to take the powerful basic science that lies there to define "complexity 2.0". And to deliver on the amazing potential that the concept of studying complexity has for science.

# Today We Put a Prize on a Small Turing Machine

*Published May 14, 2007*

It is perhaps ironic that two weeks after releasing what is probably the single most complex computational system ever constructed [Mathematica 6], we are today announcing a prize for the very simplest of computational systems.

But today is the fifth anniversary of the publication of *A New Kind of Science*, and to commemorate this, we have decided to establish the first NKS prize. The prize is related to a core objective of the basic science of NKS: to map out the abstract universe of possible computational systems.

We know from NKS that very simple programs can show immensely complex behavior. And in the NKS book I formulated the Principle of Computational Equivalence that gives an explanation for this discovery.

That principle has many predictions. And one of them is that the ability to do general-purpose computing—to be capable of universal computation—should be common even among systems with very simple rules.

Today's CPUs have millions of components. But the Principle of Computational Equivalence implies that all kinds of vastly simpler systems should also support universal computation.

The NKS book already gives several dramatic examples. But the purpose of the prize is to determine the boundary of universal computation for a particularly classic type of computational system: Turing machines.

Invented by Alan Turing in 1936 as the first successful abstract models of computation, Turing machines have been widely used in theoretical computer science. But until the whole framework of NKS, looking at specific simple Turing machines seemed merely to be a curiosity. And indeed, even though he certainly had the capabilities to do it, I do not believe that Alan Turing ever himself actually explicitly simulated any Turing machine at all.

In the NKS book, I found what is currently the simplest known universal Turing machine—with 2 states and 5 colors. I also did an extensive search of simpler Turing machines. And in doing that I found a much simpler candidate for universality: the following 2,3 Turing machine:

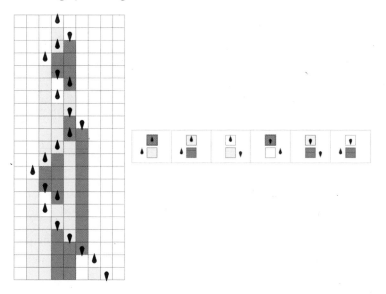

I did quite a bit of analysis on this machine. But I wasn't able to show either that it was universal, or that it wasn't. It's really easy to simulate this in Mathematica. In fact, in Mathematica 6 we just added a built-in Turing machine function. So running this machine for $t$ steps is just:

TuringMachine[{596440, 2, 3}, {1, {{}, 0}}, t]

Over the last five years, we have done a variety of kinds of things to help support the growing NKS community. Today we are adding one more: we are establishing our first NKS prize. We're offering $25,000 to the first person or group who can determine whether the 2,3 Turing machine above is actually universal, or not—and can provide the proof. All the details are at www.wolframprize.org.

# The Prize Is Won;
# The Simplest Universal Turing Machine Is Proved

*Published October 24, 2007*

"And although it will no doubt be very difficult to prove, it seems likely that this Turing machine will in the end turn out to be universal."

So I wrote on page 709 of *A New Kind of Science* (NKS). I had searched the computational universe for the simplest possible universal Turing machine. And I had found a candidate—that my intuition told me was likely to be universal. But I was not sure.

And so as part of commemorating the fifth anniversary of *A New Kind of Science* on May 14 this year, we announced a $25,000 prize for determining whether or not that Turing machine is in fact universal.

I had no idea how long it would take before the prize was won. A month? A year? A decade? A century? Perhaps the question was even formally undecidable (say from the usual axioms of mathematics). But today I am thrilled to be able to announce that after only five months the prize is won—and we have the answer: the Turing machine *is* in fact universal!

Alex Smith—a 20-year-old undergraduate from Birmingham, UK—has produced a 40-page proof. I'm pleased that my intuition was correct. But more importantly, we now have another piece of evidence for the very general Principle of Computational Equivalence (PCE) that I introduced in *A New Kind of Science*. We are also at the end of a quest that has spanned more than half a century to find the very simplest universal Turing machine.

Here it is. Just two states and three colors. And able to do any computation that can be done.

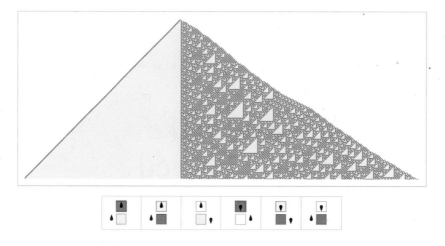

We've come a long way since Alan Turing's original 1936 universal Turing machine—taking four pages of dense notation to describe. There were some simpler universal Turing machines constructed in the mid-1900s—the record being a 7-state, 4-color machine from 1962.

That record stood for 40 years—until in 2002 I gave a 2,5 universal machine in *A New Kind of Science*. We know that no 2,2 machine can be universal. So the simplest possibility is 2,3. And from searching the 2,985,984 possible 2,3 machines, I found a candidate. Which as of today we know actually is universal.

From our everyday experience with computers, this seems pretty surprising. After all, we're used to computers whose CPUs have been carefully engineered, with millions of gates.

It seems bizarre that we should be able to achieve universal computation with a machine as simple as the one above—that we can find just by doing a little searching in the space of possible machines. But that's the new intuition that we get from NKS. That in the computational universe, phenomena like universality are actually quite common—even among systems with very simple rules.

It's just that in our normal efforts of engineering, we've been too constrained to see with such things. From all my investigation of the computational universe, I came up with the very general principle that I call the Principle of Computational Equivalence.

What it says is essentially this: that when one sees behavior that isn't obviously simple, it'll essentially always correspond to a computation that's in a sense maximally sophisticated.

One might think that computational ability would be a more gradual phenomenon: that as one increased the complexity of the rules for a system, the system would gradually show greater computational ability. But PCE says that's not how it works. It says that above a very low threshold, all systems will be exactly equivalent in their computational capabilities.

And if that's true, it has some pretty fundamental consequences. About the limits of exact science. About the occurrence of intelligence in the universe. About the phenomenon of free will. About why mathematics is difficult. About new directions in technology.

But is PCE true? I'm sure it is. But—like many fundamental principles in science—it's not the kind of thing one can abstractly prove. Instead, one has to figure out whether it's true by accumulating evidence—in effect, by doing experiments, and seeing how they come out.

Well, one of the predictions of PCE is that as soon as one sees something like a Turing machine whose behavior is complex, the system will end up being universal—even if its underlying rules are really simple.

And that's a prediction one can in principle test. The first big test was in the NKS book: that among the very simplest cellular automata, rule 110 is already universal. But what about Turing machines? Well, that's what the prize was about. And as of today we know that the prediction of PCE also checks out there.

We did an experiment; and PCE was validated. But unlike some science experiments, it didn't take a multibillion-dollar particle accelerator. It just took a 20-year-old undergraduate with a PC. I really wondered what kind of person would win our prize. I gave it about even odds of being a "professional" or an "amateur". I didn't know if the proof would require fancy number theory or other mathematics, accessible only to a "professional". Or if it could be done purely in explicit "elementary" computational terms.

But at 20:53:59 GMT on Saturday, June 30—just 47 days after we announced the prize—we received a submission, with the description of the submitter given as "Alex Smith is an undergraduate studying Electronic and Computer Engineering at the University of Birmingham, UK. He has a background in mathematics and esoteric programming languages."

We looked at it. Forty pages of detailed arguments and code. It was clear that it was a serious submission. We started to analyze it. It had clearly gotten a long way. But it hadn't quite proved universality—because it still in effect required a separate universal computer to set up each program for the 2,3 machine.

On August 1 we sent back detailed comments. And six days later a revised proof arrived—that got much closer. We sent copies to our prize committee. One of the members of the committee asked whether the proof really satisfied the formal definition of universality that he'd given in a seminal paper in 1956.

And a few weeks later we received another version clarifying that point. There's quite a bit of subtlety. Early definitions of universality assumed that programs for a Turing machine must involve only a finite number of "nonzero bits"—and that the Turing machine must "halt". But the 2,3 Turing machine—like modern computers (or systems in nature)—doesn't "halt". And in Alex Smith's construction the Turing machine "tape" (i.e., memory) must be filled with an infinite pattern of bits.

But the key point is that the pattern of bits can be set up without doing universal computation. So that means that when we see universal computation, it's really being done by the 2,3 machine, not somehow by the encoding we're using.

What Alex Smith needed to do to establish universality and win the prize was just to show that there's *some* way of programming the 2,3 machine to do any computation. That it's possible to make a "compiler" that compiles "code" for some known class of universal machines to code for the 2,3 machine.

He did that. But his "compiler" doesn't make terribly compact or efficient code. In fact, for anything but the simplest cases, the code tends to be astronomically large and horrendously inefficient. But that isn't the point here. The point is one of principle: the 2,3 Turing machine is universal. No doubt it'll be possible to find much better compilers, that make much better code.

And that'll be interesting. Perhaps one day there'll even be practical molecular computers built from this very 2,3 Turing machine. With tapes a bit like RNA strands, and heads moving up and down like ribosomes. When we think of nanoscale computers, we usually imagine carefully engineering them to mimic the architecture of the computers we know today.

But one of the lessons of NKS—brought home again by Alex Smith's proof—is that there's a completely different way to operate.

We don't have to carefully build things up with engineering. We can just go out and search in the computational universe, and find things like universal computers—that are simple enough that we can imagine making them out of molecules.

I telephoned Alex Smith a few days ago, to tell him that we were finally convinced that he'd solved the problem and earned the prize.

I asked him why he'd worked on it. He said he'd seen it as a nice puzzle. That at first he was pretty sure the Turing machine's behavior was simple enough that he could prove that it *wasn't* universal. But then, as he studied it, he realized that there were little bits of behavior that were more complicated. And it was with these that he managed to show universality.

It's a thoroughly nice piece of NKS work. Establishing a wonderful monument in the computational universe—a marker at the edge of universality for Turing machines.

We plan an official prize ceremony in a few weeks—fittingly enough, at Bletchley Park, where Alan Turing did his wartime work. But for now I'm just thrilled to see such a nice piece of science come out.

It's a very satisfying way to spend $25,000.

## Simple Turing Machines, Universality, Encodings, etc.

The following are some remarks that I made on the Foundations of Math (FOM) mailing list in connection with discussion of the Wolfram 2,3 Turing Machine Prize.

Several people forwarded me the thread on this mailing list about our 2,3 Turing Machine Prize. I'm glad to see that it has stimulated discussion. Perhaps I can make a few general remarks.

### *What do we learn from simple universal Turing machines?*

John McCarthy wrote:

> In the 1950s I thought that the smallest possible (symbol-state product) universal Turing machine would tell something about the nature of computation. Unfortunately, it didn't.

I suspect that what was imagined at that time was that by finding the smallest universal machines one would discover some "spark of computation"—some critical ingredient in the rules necessary to make universal computation possible. (At the time, it probably also still seemed that there might be a "spark of life" or a "spark of intelligence" that could be found in systems.)

I remember that when I first heard about universality in the Game of Life in the early 1970s, I didn't think it particularly significant; it seemed like just a clever hack.

But a few decades later—after building up the whole framework in *A New Kind of Science*, I have a quite different view. I now think that it's extremely significant just how widespread—or not—computational ability is. Are typical systems that we encounter in nature universal? Or are they computationally much simpler?

Can we expect to find non-trivial computational behavior just by searching a space of possible systems? Or can we only reach such behavior by specifically setting it up, say using traditional engineering methods?

I think these kinds of questions are very important for the future of both science and technology. In science, for example, they tell us to what extent we can expect to find "computationally reduced" models for systems we study. And in technology, they tell us to what extent we can expect to find interesting practical algorithms and other things just by searching the "computational universe".

I don't think that the details of which particular Turing machine is or is not universal are all that significant. What is significant, though, is how widespread universality is in general. And currently the only way we have of establishing that is to do the hard work of proving the universality or non-universality of particular Turing machines.

Our standard intuition tends to be that systems with simple rules must behave in simple ways. But what we find empirically by studying the computational universe is that that just isn't true. So now the problem is to get a scientific understanding of this, and to make precise statements about it.

One thing that's come out of my efforts in this direction is what I call the Principle of Computational Equivalence. It's a fairly bold hypothesis about which there's much to say (e.g., Chapter 12 [of *A New Kind of Science*]).

But one of its predictions is that universality should be possible even in systems with very simple rules. I've been keen to test that prediction. And looking for small universal Turing machines is a good way to do it.

There are other reasons to be interested in small universal Turing machines, too.

Perhaps one can even use them as raw material for creating useful computers out of components like molecules. Alex Smith's encoding for the 2,3 Turing machine clearly isn't going to be practical (and wasn't intended to be). But one of the lessons of modern technology is that once one knows something is in principle possible, it tends to be just a matter of time before it can be reduced to practice.

There are also theoretical computer science reasons to be interested in small universal machines. An example is understanding the tradeoffs between algorithmic complexity and computational complexity.

(For example, I made some progress in empirical studies of traditional computational complexity questions by looking at small Turing machines in Section 12.8 [of *A New Kind of Science*].)

But most of all, what I think is important about finding small universal Turing machines is that it helps build our intuition about what kinds of systems are universal—so that when we encounter systems in computer science, mathematics, natural science, or whatever, we have a better chance of knowing whether or not they're likely to be universal (and thus for example to show undecidability).

### *How should one set up the definition of universality?*

As for any concept, one wants a definition that's useful and robust, and that one can build on. Obviously the initial conditions can't need a universal computer to set them up. But just what can they contain?

I think the "obvious answer" has changed since the 1950s. In the 1950s (which were firmly before my time), my impression is that one imagined that a computer memory—or a Turing machine tape—would somehow always start with 0s in every one of its cells. (Actually, even then, I suspect that when one powered up a computer, there could be random data in its memory, which would need to be explicitly cleared.)

Now, particularly when we look at the correspondence with systems in nature, it doesn't seem so especially natural to have an infinite sequence specifically of 0s. Why not have 1s? Or 01 blocks? (Like one has digit sequences of integers vs. rationals, etc.)

I must say that I would consider it most elegant to have universality established with an initial condition that is basically just repetitive. But what is so special about repetitiveness? There are lots of other patterns one can imagine. And it seems as if the only robust distinction one can make is whether the pattern itself requires universal computation to set up.

There is no doubt some nice theoretical computer science that can be done in tightening up all these notions.

Perhaps there is really a whole hierarchy of "levels of universality"—with systems requiring different levels of computation in the preparation of their inputs to achieve universality. (One might imagine block-universal vs. finite-automaton-universal vs. context-free-universal, etc.)

My own intuition is that while there may be boundary cases where universality is possible with some forms of input but not others, this will be rare—and that most of the time a system will either be "fully universal" or not universal at all.

There may well be cases where special forms of input prevent universality. It might even be that infinite sequences of 0s prevent universality in the "prize" 2,3 Turing machine (which has rather special behavior in a field of 0s). But I'm guessing that if one considers more robust classes of encodings, it will usually matter very little which class one is using.

I'm guessing that the situation is similar to intermediate degrees. That there are in principle systems that show undecidability but not universality, but that this is extremely rare.

It would, of course, be quite wonderful to have a simple Turing machine or other system that fundamentally shows undecidability but not universality. Though I wonder slightly whether this is in any robust sense possible. After all, if one looks at the existing examples of intermediate degrees, they always seem to have "universality inside".

OK, so what about the 2,3 "prize" Turing machine? Alex Smith's proof is about universality with initial conditions created by fairly complicated (but non-universal) computations. Can it be extended to simpler initial conditions? I certainly expect the initial conditions can be considerably simpler.

Can they be a finite region embedded in an infinite sea of 0s? Obviously I don't know. Though it would be interesting to find out.

I might mention, by the way, that there are other 2,3 (and 3,2) Turing machines that I suspect are universal too. And some of them definitely have quite different behaviors with respect to infinite sequences of 0s.

### Will the encoding always be more complex if the machine is simpler?

Our standard intuition always tends to be that the more complex a thing we want to get out, the more complex a thing we have to put in. But one of the big points of *A New Kind of Science* (encapsulated, for example, in the Principle of Computational Equivalence) is that this isn't generally true.

Alex Smith's encoding for the 2,3 Turing machine is complicated and inefficient. But is this inevitable when one has a simple underlying system? I don't think so. But so far I don't really have strong evidence.

Here's one slightly related observation, though. One of the things I did in *A New Kind of Science* (and thought was rather nice) was finding the very simplest possible (equational) axiom system for Boolean algebra. It turns out to be just one axiom: $((b.c).a).(b.(b.a).b)) == a$. (See Section 12.9 [of *A New Kind of Science*].)

Now the question is: will proofs that are based on this tiny axiom system inevitably be systematically longer than ones based on bigger ("engineered") axiom systems? (Do they for example always first have to prove the axioms of the bigger axiom systems as theorems, and then go on from there?)

Well, what I found empirically is that bigger axioms aren't necessarily more efficient. (This may not be a robust result; it could depend on details of theorem-proving algorithms.)

In fact, as "Proof lengths in logic" [*A New Kind of Science*, page 1175] shows, even the single-axiom axiom system isn't terribly inefficient—at least after it proves commutativity. And the minimal axiom system I found that explicitly includes commutativity—$\{(a.b).(a.(b.c)) == a, a.b == b.a\}$—seems to be pretty much as efficient as anything. This isn't directly relevant to encodings for universality, but perhaps it gives some indications.

Another way I tried to get evidence about encodings was to see how far I could get in covering a space of possible finite cellular automaton computations by using encodings based on blocks of different lengths.

It was encouraging that cellular automata I didn't think were universal didn't manage to cover much, but ones I did think were universal did.

I looked a little at whether more complicated rules would allow smaller sets of blocks ("smaller encodings") to cover a given region of cellular automaton computation space. And didn't find any evidence of it.

I'm not quite sure in general how to formulate the problem of complexity of encodings. After all, there are inevitably computations that are arbitrarily difficult to encode for a given system.

But somehow the question is whether "reasonable" or "interesting" computations can be encoded in a short way in a given system.

It's a little like asking what it takes for a programming language to let you write the programs you want in a short way. For a language designer like me, this is an important practical question. And I'd love to have a better theoretical formulation of it. (Our "open code" Demonstrations Project has some pretty interesting examples of what can and occasionally can't be done with short Mathematica programs...)

### What happens after the prize, etc.?

I've obviously spent a huge amount of effort pursuing my interests in NKS and writing up what I've done. As well as being fascinating and fun for me, I think it's important stuff—and I've been keen to get other people involved.

We've had a lot of success with things like our Summer School. But with the prize we wanted to try stimulating the field in a different way. And so far, the experiment of the prize seems very successful.

Is the prize the end of the road for this particular 2,3 Turing machine? Definitely not. There is lots more to be studied and established about it. Perhaps it'll be proved that it can't be universal in certain senses. Probably there'll be much simpler proofs of universality found. Perhaps someone will "reduce to number theory" or some such the detailed behavior from a blank tape. Even though it has such simple rules, it's obviously a rich system to study.

And having seen how this prize has gone, I'm now motivated to think about setting up some other prizes...

# Announcing the Rule 30 Prizes

*Published October 1, 2019*

### The Story of Rule 30

How can something that simple produce something that complex? It's been nearly 40 years since I first saw rule 30—but it still amazes me. Long ago it became my personal all-time favorite science discovery, and over the years it's changed my whole worldview and led me to all sorts of science, technology, philosophy and more.

But even after all these years, there are still many basic things we don't know about rule 30. And I've decided that it's now time to do what I can to stimulate the process of finding more of them out. So as of today, I am offering $30,000 in prizes for the answers to three basic questions about rule 30.

The setup for rule 30 is extremely simple. One's dealing with a sequence of lines of black and white cells. And given a particular line of black and white cells, the colors of the cells on the line below are determined by looking at each cell and its immediate neighbors and then applying the following simple rule:

If you start with a single black cell, what will happen? One might assume—as I at first did—that the rule is simple enough that the pattern it produces must somehow be correspondingly simple. But if you actually do the experiment, here's what you find happens over the first 50 steps:

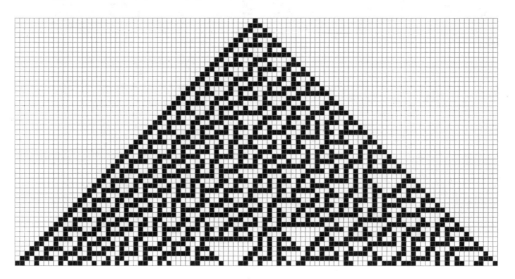

But surely, one might think, this must eventually resolve into something much simpler. Yet here's what happens over the first 300 steps:

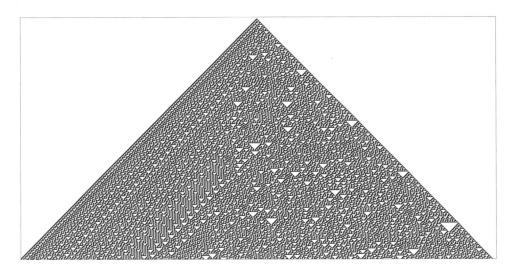

And, yes, there's some regularity over on the left. But many aspects of this pattern look for all practical purposes random. It's amazing that a rule so simple can produce behavior that's so complex. But I've discovered that in the computational universe of possible programs this kind of thing is common, even ubiquitous. And I've built a whole new kind of science—with all sorts of principles—based on this.

And gradually there's been more and more evidence for these principles. But what specifically can rule 30 tell us? What concretely can we say about how it behaves? Even the most obvious questions turn out to be difficult. And after decades without answers, I've decided it's time to define some specific questions about rule 30, and offer substantial prizes for their solutions.

I did something similar in 2007, putting a prize on a core question about a particular Turing machine. And at least in that case the outcome was excellent. In just a few months, the prize was won—establishing forever what the simplest possible universal Turing machine is, as well as providing strong further evidence for my general Principle of Computational Equivalence.

The Rule 30 Prize Problems again get at a core issue: just how complex really is the behavior of rule 30? Each of the problems asks this in a different, concrete way. Like rule 30 itself, they're all deceptively simple to state. Yet to solve any of them will be a major achievement—that will help illuminate fundamental principles about the computational universe that go far beyond the specifics of rule 30.

I've wondered about every one of the problems for more than 35 years. And all that time I've been waiting for the right idea, or the right kind of mathematical or computational thinking, to finally be able to crack even one of them. But now I want to open this process up to the world. And I'm keen to see just what can be achieved, and what methods it will take.

## The Rule 30 Prize Problems

For the Rule 30 Prize Problems, I'm concentrating on a particularly dramatic feature of rule 30: the apparent randomness of its center column of cells. Start from a single black cell, then just look down the sequence of values of this cell—and it seems random:

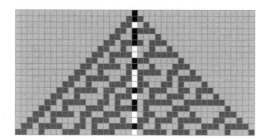

But in what sense is it really random? And can one prove it? Each of the Prize Problems in effect uses a different criterion for randomness, then asks whether the sequence is random according to that criterion.

### Problem 1: Does the center column always remain non-periodic?

Here's the beginning of the center column of rule 30:

▬▬ ▬▬▬▬ ▬▬ ▬ ▬ ▬ ▬ ▬ ▬▬▬ ▬ ▬▬▬ ▬▬▬ ▬ ▬ ▬▬ ▬▬ ▬ ▬ ▬▬ ▬ ▬ ▬▬▬▬▬ ▬▬▬ ▬

It's easy to see that this doesn't repeat—it doesn't become periodic. But this problem is about whether the center column ever becomes periodic, even after an arbitrarily large number of steps. Just by running rule 30, we know the sequence doesn't become periodic in the first billion steps. But what about ever? To establish that, we need a proof. (Here are the first million (wolfr.am/A-Million-Bits) and first billion (wolfr.am/A-Billion-Bits) bits in the sequence, by the way, as entries in the Wolfram Data Repository (wolfr.am/datarepository).)

**Problem 2: Does each color of cell occur on average equally often in the center column?**

Here's what one gets if one tallies the number of black and of white cells in successively more steps in the center column of rule 30:

| steps | ■ | □ | ratio |
|---|---|---|---|
| 1 | 1 | 0 | |
| 10 | 7 | 3 | 2.33333 |
| 100 | 52 | 48 | 1.08333 |
| 1000 | 481 | 519 | 0.92678 |
| 10 000 | 5032 | 4968 | 1.01288 |
| 100 000 | 50 098 | 49 902 | 1.00393 |
| 1 000 000 | 500 768 | 499 232 | 1.00308 |
| 10 000 000 | 5 002 220 | 4 997 780 | 1.00089 |
| 100 000 000 | 50 009 976 | 49 990 024 | 1.00040 |
| 1 000 000 000 | 500 025 038 | 499 974 962 | 1.00010 |

The results are certainly close to equal for black vs. white. But what this problem asks is whether the limit of the ratio after an arbitrarily large number of steps is exactly 1.

**Problem 3: Does computing the $n^{th}$ cell of the center column require at least $O(n)$ computational effort?**

To find the $n^{th}$ cell in the center column, one can always just run rule 30 for $n$ steps, computing the values of all the cells in this diamond:

But if one does this directly, one's doing $\frac{1}{2} n^2$ individual cell updates, so the computational effort required goes up like $O(n^2)$. This problem asks if there's a shortcut way to compute the value of the $n^{th}$ cell, without all this intermediate computation—or, in particular, in less than $O(n)$ computational effort.

## The Digits of π

Rule 30 is a creature of the computational universe: a system found by exploring possible simple programs with the new intellectual framework that the paradigm of computation provides. But the problems I've defined about rule 30 have analogs in mathematics that are centuries old.

Consider the digits of π. They're a little like the center column of rule 30. There's a definite algorithm for generating them. Yet once generated they seem for all practical purposes random:

3.1415926535897932384626433832795028841971693993751058209749445923078164062862086208998628

Just to make the analog a little closer, here are the first few digits of π in base 2:

$11.0010010000111111011010101000100010000101101000110000100011010011000100110001100112_2$

And here are the first few bits in the center column of rule 30:

1101110011000101100100111010111001110101011000011001010110101011111110000111100010

Just for fun, one can convert these to base 10:

0.8623897839473840486408002460867511281085329636245506152619584529179320275892 3479

Of course, the known algorithms for generating the digits of π are considerably more complicated than the simple rule for generating the center column of rule 30. But, OK, so what's known about the digits of π?

Well, we know they don't repeat. That was proved in the 1760s when it was shown that π is an irrational number—because the only numbers whose digits repeat are rational numbers. (It was also shown in 1882 that π is transcendental, i.e., that it cannot be expressed in terms of roots of polynomials.)

How about the analog of problem 2? Do we know if in the digit sequence of π different digits occur with equal frequency? By now more than 100 trillion binary digits have been computed—and the measured frequencies of digits are very close (in the first 40 trillion binary digits the ratio of 1s to 0s is about 0.9999998064). But in the limit, are the frequencies exactly the same? People have been wondering about this for several centuries. But so far mathematics hasn't succeeded in delivering any results.

For rational numbers, digit sequences are periodic, and it's easy to work out relative frequencies of digits. But for the digit sequences of all other "naturally constructed" numbers, basically there's nothing known about limiting frequencies of digits. It's a reasonable guess that actually the digits of $\pi$ (as well as the center column of rule 30) are "normal", in the sense that not only every individual digit, but also every block of digits of any given length in the limit occur with equal frequency. And as was noted in the 1930s, it's perfectly possible to "digit-construct" normal numbers. Champernowne's number, formed by concatenating the digits of successive integers, is an example (and, yes, this works in any base, and one can also get normal numbers by concatenating values of functions of successive integers):

0.12345678910111213141516171819202122232425262728293031323334353637383940414243445

But the point is that for "naturally constructed" numbers formed by combinations of standard mathematical functions, there's simply no example known where any regularity of digits has been found. Of course, it ultimately depends what one means by "regularity"—and at some level the problem devolves into a kind of number-digit analog of the search for extraterrestrial intelligence. But there's absolutely no proof that one couldn't, for example, find even some strange combination of square roots that would have a digit sequence with some very obvious regularity.

OK, so what about the analog of problem 3 for the digits of $\pi$? Unlike rule 30, where the obvious way to compute elements in the sequence is one step at a time, traditional ways of computing digits of $\pi$ involve getting better approximations to $\pi$ as a complete number. With the standard (bizarre-looking) series invented by Ramanujan in 1910 and improved by the Chudnovsky brothers in 1989, the first few terms in the series give the following approximations:

3.141592653589793238462643383587350688475866345996374315654905806801301450565203591105830910219290929
3.141592653589793238462643383279502884197167678854846287912727790370642977335176958726922911495373797
3.141592653589793238462643383279502884197169399375105820984947408020662452789717346364103622321101908
3.141592653589793238462643383279502884197169399375105820974944592307816346694690247717268165239156011
3.141592653589793238462643383279502884197169399375105820974944592307816406286208998628395732194831867
3.141592653589793238462643383279502884197169399375105820974944592307816406286208998628034825342117066
3.141592653589793238462643383279502884197169399375105820974944592307816406286208998628034825342117068
3.141592653589793238462643383279502884197169399375105820974944592307816406286208998628034825342117068
3.141592653589793238462643383279502884197169399375105820974944592307816406286208998628034825342117068

So how much computational effort is it to find the $n^{th}$ digit? The number of terms required in the series is $O(n)$. But each term needs to be computed to $n$-digit precision, which requires at least $O(n)$ individual digit operations—implying that altogether the computational effort required is more than $O(n)$.

Until the 1990s it was assumed that there wasn't any way to compute the $n^{th}$ digit of $\pi$ without computing all previous ones. But in 1995 Simon Plouffe discovered that actually it's possible to compute—albeit slightly probabilistically—the $n^{th}$ digit without computing earlier ones. And while one might have thought that this would allow the $n^{th}$ digit to be obtained with less than $O(n)$ computational effort, the fact that one has to do computations at $n$-digit precision means that at least $O(n)$ computational effort is still required.

## Results, Analogies and Intuitions

### Problem 1: Does the center column always remain non-periodic?

Of the three Rule 30 Prize Problems, this is the one on which the most progress has already been made. Because while it's not known if the center column in the rule 30 pattern ever becomes periodic, Erica Jen showed in 1986 that no two columns can both become periodic. And in fact, one can also give arguments that a single column plus scattered cells in another column can't both be periodic.

The proof about a pair of columns uses a special feature of rule 30. Consider the structure of the rule:

Normally one would just say that given each triple of cells, the rule determines the color of the center cell below. But for rule 30, one can effectively also run the rule sideways: given the cell to the right and above, one can also uniquely determine the color of the cell to the left. And what this means is that if one is given two adjacent columns, it's possible to reconstruct the whole pattern to the left:

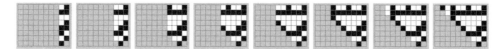

But if the columns were periodic, it immediately follows that the reconstructed pattern would also have to be periodic. Yet by construction at least the initial condition is definitely not periodic, and hence the columns cannot both be periodic. The same argument works if the columns are not adjacent, and if one doesn't know every cell in both columns. But there's no known way to extend the argument to a single column—such as the center column—and thus it doesn't resolve the first Rule 30 Prize Problem.

OK, so what would be involved in resolving it? Well, if it turns out that the center column is eventually periodic, one could just compute it, and show that. We know it's not periodic for the first billion steps, but one could at least imagine that there could be a trillion-step transient, after which it's periodic.

Is that plausible? Well, transients do happen—and theoretically (just like in the classic Turing machine halting problem) they can even be arbitrarily long. Here's a somewhat funky example—found by a search—of a rule with 4 possible colors (totalistic code 150898). Run it for 200 steps, and the center column looks quite random:

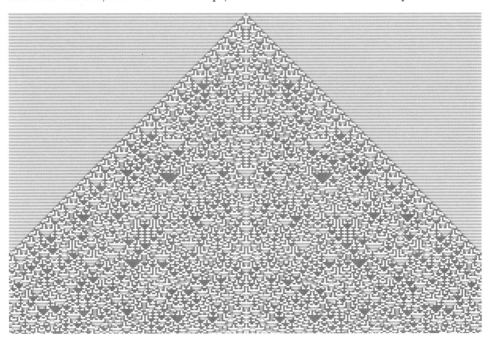

After 500 steps, the whole pattern still looks quite random:

But if one zooms in around the center column, there's something surprising: after 251 steps, the center column seems to evolve to a fixed value (or at least it's fixed for more than a million steps):

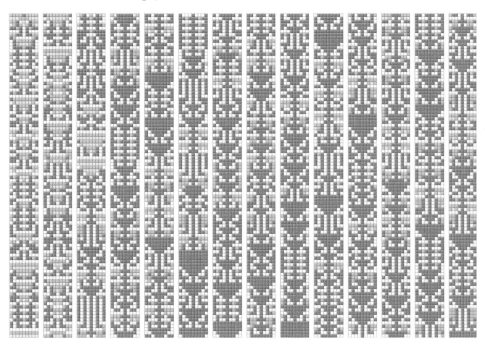

Could some transient like this happen in rule 30? Well, take a look at the rule 30 pattern, now highlighting where the diagonals on the left are periodic:

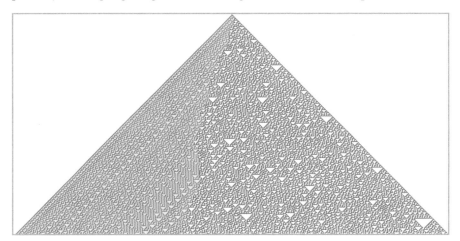

There seems to be a boundary that separates order on the left from disorder on the right. And at least over the first 100,000 or so steps, the boundary seems to move on average about 0.252 steps to the left at each step—with roughly random fluctuations:

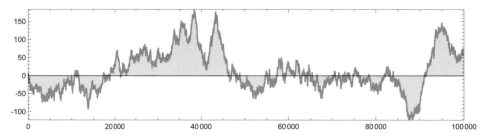

But how do we know that there won't at some point be a huge fluctuation, that makes the order on the left cross the center column, and perhaps even make the whole pattern periodic? From the data we have so far, it looks unlikely, but I don't know any way to know for sure.

And it's certainly the case that there are systems with exceptionally long "transients". Consider the distribution of primes, and compute LogIntegral[*n*]-PrimePi[*n*]:

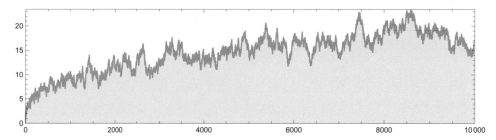

Yes, there are fluctuations. But from this picture it certainly looks as if this difference is always going to be positive. And that's, for example, what Ramanujan thought. But it turns out it isn't true. At first the bound for where it would fail was astronomically large (Skewes's number 10^10^10^964). And although still nobody has found an explicit value of $n$ for which the difference is negative, it's known that before $n = 10^{317}$ there must be one (and eventually the difference will be negative at least nearly a millionth of the time).

I strongly suspect that nothing like this happens with the center column of rule 30. But until we have a proof that it can't, who knows?

One might think, by the way, that while one might be able to prove periodicity by exposing regularity in the center column of rule 30, nothing like that would be possible for non-periodicity. But actually, there are patterns whose center columns one can readily see are non-periodic, even though they're very regular. The main class of examples are nested patterns. Here's a very simple example, from rule 161—in which the center column has white cells when $n = 2^k$:

Here's a slightly more elaborate example (from the 2-neighbor 2-color rule 69540422), in which the center column is a Thue–Morse sequence ThueMorse[$n$]:

One can think of the Thue–Morse sequence as being generated by successively applying the substitutions:

And it turns out that the $n^{\text{th}}$ term in this sequence is given by Mod[DigitCount[$n$, 2, 1], 2] —which is never periodic.

Will it turn out that the center column of rule 30 can be generated by a substitution system? Again, I'd be amazed (although there are seemingly natural examples where very complex substitution systems do appear). But once again, until one has a proof, who knows?

Here's something else, that may be confusing, or may be helpful. The Rule 30 Prize Problems all concern rule 30 running in an infinite array of cells. But what if one considers just $n$ cells, say with the periodic boundary conditions (i.e., taking the right neighbor of the rightmost cell to be the leftmost cell, and vice versa)? There are $2^n$ possible total states of the system—and one can draw a state transition diagram that shows which state evolves to which other. Here's the diagram for $n = 5$:

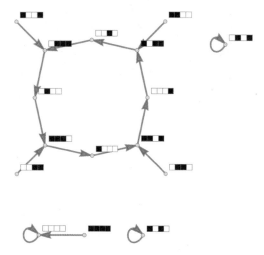

And here it is for $n = 4$ through $n = 11$:

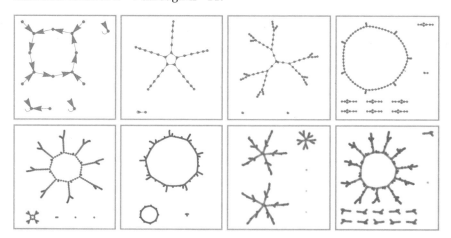

The structure is that there are a bunch of states that appear only as transients, together with other states that are on cycles. Inevitably, no cycle can be longer than $2^n$ (actually, symmetry considerations show that it always has to be somewhat less than this).

OK, so on a size-$n$ array, rule 30 always has to show behavior that becomes periodic with a period that's less than $2^n$. Here are the actual periods starting from a single black cell initial condition, plotted on a log scale:

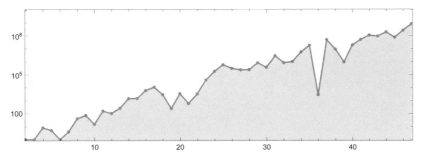

And at least for these values of $n$, a decent fit is that the period is about $2^{0.63\,n}$. And, yes, at least in all these cases, the period of the center column is equal to the period of the whole evolution. But what do these finite-size results imply about the infinite-size case? I, at least, don't immediately see.

**Problem 2: Does each color of cell occur on average equally often in the center column?**
Here's a plot of the running excess of 1s over 0s in 10,000 steps of the center column of rule 30:

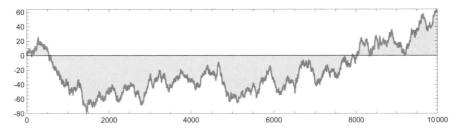

Here it is for a million steps:

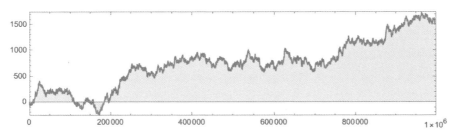

And a billion steps:

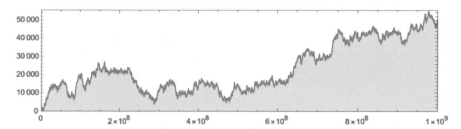

We can see that there are times when there's an excess of 1s over 0s, and vice versa, though, yes, as we approach a billion steps 1 seems to be winning over 0, at least for now. But let's compute the ratio of the total number of 1s to the total number of 0s. Here's what we get after 10,000 steps:

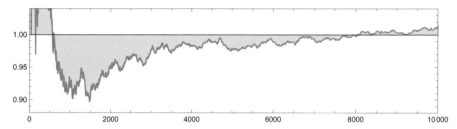

Is this approaching the value 1? It's hard to tell. Go on a little longer, and this is what we see:

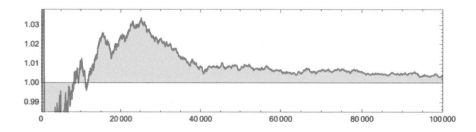

The scale is getting smaller, but it's still hard to tell what will happen. Plotting the difference from 1 on a log-log plot up to a billion steps suggests it's fairly systematically getting smaller:

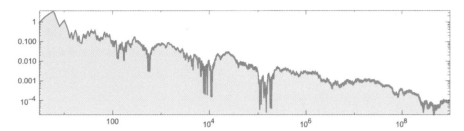

But how do we know this trend will continue? Right now, we don't. And, actually, things could get quite pathological. Maybe the fluctuations in 1s vs. 0s grow, so even though we're averaging over longer and longer sequences, the overall ratio will never converge to a definite value.

Again, I doubt this is going to happen in the center column of rule 30. But without a proof, we don't know for sure.

We're asking here about the frequencies of black and white cells. But an obvious—and potentially illuminating—generalization is to ask instead about the frequencies for blocks of cells of length $k$. We can ask if all $2^k$ such blocks have equal limiting frequency. Or we can ask the more basic question of whether all the blocks even ever occur—or, in other words, whether if one goes far enough, the center column of rule 30 will contain any given sequence of length $k$ (say a bitwise representation of some work of literature).

Again, we can get empirical evidence. For example, at least up to $k = 22$, all $2^k$ sequences do occur—and here's how many steps it takes:

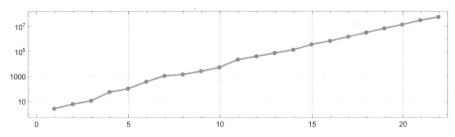

It's worth noticing that one can succeed perfectly for blocks of one length, but then fail for larger blocks. For example, the Thue–Morse sequence mentioned above has exactly equal frequencies of 0 and 1, but pairs don't occur with equal frequencies, and triples of identical elements simply never occur.

In traditional mathematics—and particularly dynamical systems theory—one approach to take is to consider not just evolution from a single-cell initial condition, but evolution from all possible initial conditions. And in this case it's straightforward to show that, yes, if one evolves with equal probability from all possible initial conditions, then columns of cells generated by rule 30 will indeed contain every block with equal frequency. But if one asks the same thing for different distributions of initial conditions, one gets different results, and it's not clear what the implication of this kind of analysis is for the specific case of a single-cell initial condition.

If different blocks occurred with different frequencies in the center column of rule 30, then that would immediately show that the center column is "not random", or in other words that it has statistical regularities that could be used to at least statistically predict it. Of course, at some level the center column is completely "predictable": you just have to run rule 30 to find it. But the question is whether, given just the values in the center column on their own, there's a way to predict or compress them, say with much less computational effort than generating an arbitrary number of steps in the whole rule 30 pattern.

One could imagine running various data compression or statistical analysis algorithms, and asking whether they would succeed in finding regularities in the sequence. And particularly when one starts thinking about the overall computational capabilities of rule 30, it's conceivable that one could prove something about how across a spectrum of possible analysis algorithms, there's a limit to how much they could "reduce" the computation associated with the evolution of rule 30. But even given this, it'd likely still be a major challenge to say anything about the specific case of relative frequencies of black and white cells.

It's perhaps worth mentioning one additional mathematical analog. Consider treating the values in a row of the rule 30 pattern as digits in a real number, say with the first digit of the fractional part being on the center column. Now, so far as we know, the evolution of rule 30 has no relation to any standard operations (like multiplication or taking powers) that one does on real numbers. But we can still ask about the sequence of numbers formed by looking at the right-hand side of the rule 30 pattern. Here's a plot for the first 200 steps:

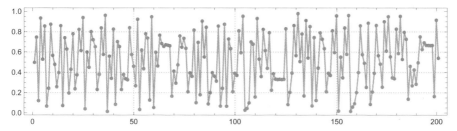

And here's a histogram of the values reached at successively more steps:

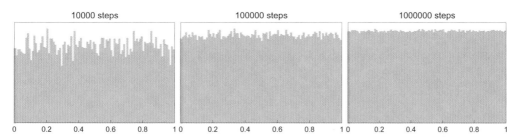

And, yes, it's consistent with the limiting histogram being flat, or in other words, with these numbers being uniformly distributed in the interval 0 to 1.

Well, it turns out that in the early 1900s there were a bunch of mathematical results established about this kind of equidistribution. In particular, it's known that FractionalPart[$h^n$] for successive $n$ is always equidistributed if $h$ isn't a rational number. It's also known that FractionalPart[$h^n$] is equidistributed for almost all $h$ (Pisot numbers like the golden ratio are exceptions). But specific cases—like FractionalPart[$(3/2)^n$]— have eluded analysis for at least half a century. (By the way, it's known that the digits of $\pi$ in base 16 and thus base 2 can be generated by a recurrence of the form $x_n$ = FractionalPart[$16\,x_{n-1} + r[n]$] where $r[n]$ is a fixed rational function of $n$.)

**Problem 3: Does computing the $n^{th}$ cell of the center column require at least O($n$) computational effort?**

Consider the pattern made by rule 150:

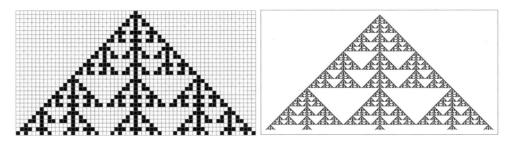

It's a very regular, nested pattern. Its center column happens to be trivial (all cells are black). But if we look one column to the left or right, we find:

How do we work out the value of the $n^{th}$ cell? Well, in this particular case, it turns out there's essentially just a simple formula: the value is given by Mod[IntegerExponent[$n$, 2], 2]. In other words, just look at the number $n$ in base 2, and ask whether the number of zeros it has at the end is even or odd.

How much computational effort does it take to "evaluate this formula"? Well, even if we have to check every bit in $n$, there are only about Log[2, $n$] of those. So we can expect that the computational effort is O(log $n$).

But what about the rule 30 case? We know we can work out the value of the $n^{th}$ cell in the center column just by explicitly applying the rule 30 update rule $n^2$ times. But the question is whether there's a way to reduce the computational work that's needed. In the past, there's tended to be an implicit assumption throughout the mathematical sciences that if one has the right model for something, then by just being clever enough one will always find a way to make predictions—or in other words, to work out what a system will do, using a lot less computational effort than the actual evolution of the system requires.

And, yes, there are plenty of examples of "exact solutions" (think 2-body problem, 2D Ising model, etc.) where we essentially just get a formula for what a system will do. But there are also other cases (think 3-body problem, 3D Ising model, etc.) where this has never successfully been done.

And as I first discussed in the early 1980s, I suspect that there are actually many systems (including these) that are computationally irreducible, in the sense that there's no way to significantly reduce the amount of computational work needed to determine their behavior.

So in effect Problem 3 is asking about the computational irreducibility of rule 30—or at least a specific aspect of it. (The choice of $O(n)$ computational effort is somewhat arbitrary; another version of this problem could ask for $O(n^\alpha)$ for any $\alpha < 2$, or, for that matter, $O(\log^\beta (n))$—or some criterion based on both time and memory resources.)

If the answer to Problem 3 is negative, then the obvious way to show this would just be to give an explicit program that successfully computes the $n^{\text{th}}$ value in the center column with less than $O(n)$ computational effort, as we did for rule 150 above.

We can ask what $O(n)$ computational effort means. What kind of system are we supposed to use to do the computation? And how do we measure "computational effort"? The phenomenon of computational universality implies that—within some basic constraints—it ultimately doesn't matter.

For definiteness we could say that we always want to do the computation on a Turing machine. And for example we can say that we'll feed the digits of the number $n$ in as the initial state of the Turing machine tape, then expect the Turing machine to grind for much less than $n$ steps before generating the answer (and, if it's really to be "formula like", more like $O(\log n)$ steps).

We don't need to base things on a Turing machine, of course. We could use any kind of system capable of universal computation, including a cellular automaton, and, for that matter, the whole Wolfram Language. It gets a little harder to measure "computational effort" in these systems. Presumably in a cellular automaton we'd want to count the total number of cell updates done. And in the Wolfram Language we might end up just actually measuring CPU time for executing whatever program we've set up.

I strongly suspect that rule 30 is computationally irreducible, and that Problem 3 has an affirmative answer. But if isn't, my guess is that eventually there'll turn out to be a program that rather obviously computes the $n^{\text{th}}$ value in less than $O(n)$ computational effort, and there won't be a lot of argument about the details of whether the computational resources are counted correctly.

But proving that no such program exists is a much more difficult proposition. And even though I suspect computational irreducibility is quite ubiquitous, it's always very hard to prove explicit lower bounds on the difficulty of doing particular computations. And in fact almost all explicit lower bounds currently known are quite weak, and essentially boil down just to arguments about information content—like that you need $O(\log n)$ steps to even read all the digits in the value of $n$.

Undoubtedly the most famous lower-bound problem is the P vs. NP question. I don't think there's a direct relation to our rule 30 problem (which is more like a P vs. LOG-TIME question), but it's perhaps worth understanding how things are connected. The basic point is that the forward evolution of a cellular automaton, say for $n$ steps from an initial condition with $n$ cells specified, is at most an $O(n^2)$ computation, and is therefore in P ("polynomial time"). But the question of whether there exists an initial condition that evolves to produce some particular final result is in NP. If you happen ("non-deterministically") to pick the correct initial condition, then it's polynomial time to check that it's correct. But there are potentially $2^n$ possible initial conditions to check.

Of course there are plenty of cellular automata where you don't have to check all these $2^n$ initial conditions, and a polynomial-time computation clearly suffices. But it's possible to construct a cellular automaton where finding the initial condition is an NP-complete problem, or in other words, where it's possible to encode any problem in NP in this particular cellular automaton inversion problem. Is the rule 30 inversion problem NP-complete? We don't know, though it seems conceivable that it could be proved to be (and if one did prove it then rule 30 could finally be a provably NP-complete cryptosystem).

But there doesn't seem to be a direct connection between the inversion problem for rule 30, and the problem of predicting the center column. Still, there's at least a more direct connection to another global question: whether rule 30 is computation universal, or, in other words, whether there exist initial conditions for rule 30 that allow it to be "programmed" to perform any computation that, for example, any Turing machine can perform.

We know that among the 256 simplest cellular automata, rule 110 is universal (as are three other rules that are simple transformations of it). But looking at a typical example of rule 110 evolution, it's already clear that there are definite, modular structures one can identify. And indeed the proof proceeds by showing how one can "engineer" a known universal system out of rule 110 by appropriately assembling these structures.

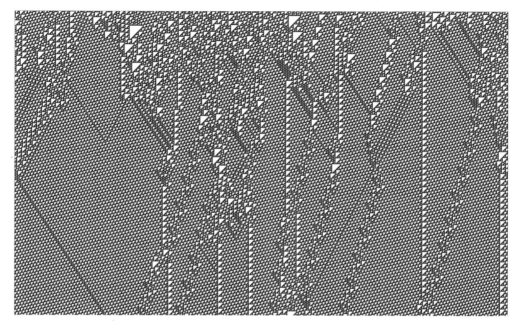

Rule 30, however, shows no such obvious modularity—so it doesn't seem plausible that one can establish universality in the "engineering" way it's been established for all other known-to-be-universal systems. Still, my Principle of Computational Equivalence strongly suggests that rule 30 is indeed universal; we just don't yet have an obvious direction to take in trying to prove it.

If one can show that a system is universal, however, then this does have implications that are closer to our rule 30 problem. In particular, if a system is universal, then there'll be questions (like the halting problem) about its infinite-time behavior that will be undecidable, and which no guaranteed-finite-time computation can answer. But as such, universality is a statement about the existence of initial conditions that reproduce a given computation. It doesn't say anything about the specifics of a particular initial condition—or about how long it will take to compute a particular result.

OK, but what about a different direction: what about getting empirical evidence about our Problem 3? Is there a way to use statistics, or cryptanalysis, or mathematics, or machine learning to even slightly reduce the computational effort needed to compute the $n^{\text{th}}$ value in the center column?

Well, we know that the whole 2D pattern of rule 30 is far from random. In fact, of all $2^{m^2}$ patches, only $m \times m$ can possibly occur—and in practice the number weighted by probability is much smaller. And I don't doubt that facts like this can be used to reduce the effort to compute the center column to less than $O(n^2)$ effort (and that would be a nice partial result). But can it be less than $O(n)$ effort? That's a much more difficult question.

Clearly if Problem 1 was answered in the negative then it could be. But in a sense asking for less than $O(n)$ computation of the center column is precisely like asking whether there are "predictable regularities" in it. Of course, even if one could find small-scale statistical regularities in the sequence (as answering Problem 2 in the negative would imply), these wouldn't on their own give one a way to do more than perhaps slightly improve a constant multiplier in the speed of computing the sequence.

Could there be some systematically reduced way to compute the sequence using a neural net—which is essentially a collection of nested real-number functions? I've tried to find such a neural net using our current deep-learning technology—and haven't been able to get anywhere at all.

What about statistical methods? If we could find statistical non-randomness in the sequence, then that would imply an ability to compress the sequence, and thus some redundancy or predictability in the sequence. But I've tried all sorts of statistical randomness tests on the center column of rule 30—and never found any significant deviation from randomness. (And for many years—until we found a slightly more efficient rule—we used sequences from finite-size rule 30 systems as our source of random numbers in the Wolfram Language, and no legitimate "it's not random!" bugs ever showed up.)

Statistical tests of randomness typically work by saying, "Take the supposedly random sequence and process it in some way, then see if the result is obviously non-random". But what kind of processing should be done? One might see if blocks occur with equal frequency, or if correlations exist, or if some compression algorithm succeeds in doing compression. But typically batteries of tests end up seeming a bit haphazard and arbitrary. In principle one can imagine enumerating all possible tests—by enumerating all possible programs that can be applied to the sequence. But I've tried doing this, for example for classes of cellular automaton rules—and have never managed to detect any non-randomness in the rule 30 sequence.

So how about using ideas from mathematics to predict the rule 30 sequence? Well, as such, rule 30 doesn't seem connected to any well-developed area of math. But of course it's conceivable that some mapping could be found between rule 30 and ideas, say, in an area like number theory—and that these could either help in finding a shortcut for computing rule 30, or could show that computing it is equivalent to some problem like integer factoring that's thought to be fundamentally difficult.

I know a few examples of interesting interplays between traditional mathematical structures and cellular automata. For example, consider the digits of successive powers of 3 in base 2 and in base 6:

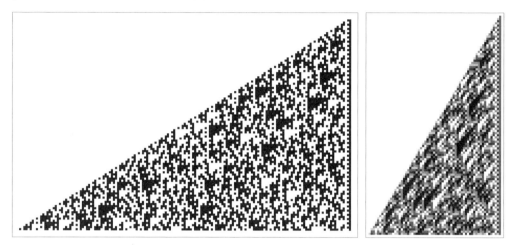

It turns out that in the base 6 case, the rule for generating the pattern is exactly a cellular automaton. (For base 2, there are additional long-range carries.) But although both these patterns look complex, it turns out that their mathematical structure lets us speed up making certain predictions about them.

Consider the $s^{th}$ digit from the right-hand edge of line $n$ in each pattern. It's just the $s^{th}$ digit in $3^n$, which is given by the "formula" (where $b$ is the base, here 2 or 6) Mod[Quotient[$3^n$, $b^s$], b]. But how easy is it to evaluate this formula? One might think that to compute $3^n$ one would have to do $n$ multiplications. But this isn't the case: instead, one can for example build up $3^n$ using repeated squaring, with about log($n$) multiplications. That this is possible is a consequence of the associativity of multiplication. There's nothing obviously like that for rule 30—but it's always conceivable that some mapping to a mathematical structure like this could be found.

Talking of mathematical structure, it's worth mentioning that there are more formula-like ways to state the basic rule for rule 30. For example, taking the values of three adjacent cells to be $p$, $q$, $r$ the basic rule is just $p \veebar (q \vee r)$, or Xor[p, Or[q, r]].

With numerical cell values 0 and 1, the basic rule is just $\text{Mod}[p + q + r + qr, 2]$. Do these forms help? I don't know. But, for example, it's remarkable that in a sense all the complexity of rule 30 comes from the presence of that one little nonlinear $q\,r$ term—for without that term, one would have rule 150, about which one can develop a complete algebraic theory using quite traditional mathematics.

To work out $n$ steps in the evolution of rule 30, one's effectively got to repeatedly compose the basic rule. And so far as one can tell, the symbolic expressions that arise just get more and more complicated—and don't show any sign of simplifying in such a way as to save computational work.

In Problem 3, we're talking about the computational effort to compute the $n^{\text{th}}$ value in the center column of rule 30—and asking if it can be less than $O(n)$. But imagine that we have a definite algorithm for doing the computation. For any given $n$, we can see what computational resources it uses. Say the result is $r[n]$. Then what we're asking is whether $r[n]$ is less than "big O" of $n$, or whether $\text{MaxLimit}[r[n]/n, n \to \infty] < \infty$.

But imagine that we have a particular Turing machine (or some other computational system) that's implementing our algorithm. It could be that $r[n]$ will at least asymptotically just be a smooth or otherwise regular function of $n$ for which it's easy to see what the limit is. But if one just starts enumerating Turing machines, one encounters examples where $r[n]$ appears to have peaks of random heights in random places. It might even be that somewhere there'd be a value of $n$ for which the Turing machine doesn't halt (or whatever) at all, so that $r[n]$ is infinite. And in general, as we'll discuss in more detail later, it could even be undecidable just how $r[n]$ grows relative to $O(n)$.

**Formal Statements of the Problems**

So far, I've mostly described the Prize Problems in words. But we can also describe them in computational language (or effectively also in math).

In the Wolfram Language, the first $t$ values in the center column of rule 30 are given by:

c[t_] := CellularAutomaton[30, {{1}, 0}, {t, {{0}}}]

And with this definition, the three problems can be stated as predicates about $c[t]$.

**Problem 1: Does the center column always remain non-periodic?**

$\nexists_{\{p,i\}} \forall_{t,t>i} \, c[t+p] == c[t]$

or

NotExists[{p, i}, ForAll[t, t > i, c[t+p] == c[t]]]

or "there does not exist a period $p$ and an initial length $i$ such that for all $t$ with $t > i$, $c[p+t]$ equals $c[t]$".

**Problem 2: Does each color of cell occur on average equally often in the center column?**

$$\lim_{\substack{t \to \infty \\ z}} \frac{\text{Total}[c[t]]}{t} == \frac{1}{2}$$

or

DiscreteLimit[Total[c[t]]/t, t → Infinity] == 1/2

or "the discrete limit of the total of the values in $c[t]/t$ as $t \to \infty$ is $1/2$".

**Problem 3: Does computing the $n^{\text{th}}$ cell of the center column require at least O(n) computational effort?**

Define machine[m] to be a machine parametrized by $m$ (for example TuringMachine[...]), and let machine[m][n] give {v, t}, where $v$ is the output value, and $t$ is the amount of computational effort taken (e.g., number of steps). Then the problem can be formulated as:

$\nexists_m (\forall_n \text{machine}[m][n][\![1]\!] == \text{Last}[c[n]] \land \varlimsup_{n \to \infty} \frac{\text{machine}[m][n][\![2]\!]}{n} < \infty)$

or "there does not exist a machine $m$ which for all $n$ gives $c[n]$, and for which the lim sup of the amount of computational effort spent, divided by $n$, is finite". (Yes, one should also require that $m$ be finite, so the machine's rule can't just store the answer.)

## The Formal Character of Solutions

Before we discuss the individual problems, an obvious question to ask is what the interdependence of the problems might be. If the answer to Problem 3 is negative (which I very strongly doubt), then it holds the possibility for simple algorithms or formulas from which the answers to Problems 1 and 2 might become straightforward. If the answer to Problem 3 is affirmative (as I strongly suspect), then it implies that the answer to Problem 1 must also be affirmative. The contrapositive is also true: if the answer to Problem 1 is negative, then it implies that the answer to Problem 3 must also be negative.

If the answer to Problem 1 is negative, so that there is some periodic sequence that appears in the center column, then if one explicitly knows that sequence, one can immediately answer Problem 2. One might think that answering Problem 2 in the negative would imply something about Problem 3. And, yes, unequal probabilities for black and white implies compression by a constant factor in a Shannon-information way. But to compute value with less than $O(n)$ resources—and therefore to answer Problem 3 in the negative—requires that one be able to identify in a sense infinitely more compression.

So what does it take to establish the answers to the problems?

If Problem 1 is answered in the negative, then one can imagine explicitly exhibiting the pattern generated by rule 30 at some known step—and being able to see the periodic sequence in the center. Of course, Problem 1 could still be answered in the negative, but less constructively. One might be able to show that eventually the sequence has to be periodic, but not know even any bound on where this might happen. If Problem 3 is answered in the negative, a way to do this is to explicitly give an algorithm (or, say, a Turing machine) that does the computation with less than $O(n)$ computational resources.

But let's say one has such an algorithm. One still has to prove that for all $n$, the algorithm will correctly reproduce the $n^{th}$ value. This might be easy. Perhaps there would just be a proof by induction or some such. But it might be arbitrarily hard. For example, it could be that for most $n$, the running time of the algorithm is clearly less than $n$. But it might not be obvious that the running time will always even be finite. Indeed, the "halting problem" for the algorithm might simply be undecidable. But just showing that a particular algorithm doesn't halt for a given $n$ doesn't really tell one anything about the answer to the problem. For that one would have to show that there's no algorithm that exists that will successfully halt in less than $O(n)$ time.

The mention of undecidability brings up an issue, however: just what axiom system is one supposed to use to answer the problems? For the purposes of the Prize, I'll just say "the traditional axioms of standard mathematics", which one can assume are Peano arithmetic and/or the axioms of set theory (with or without the continuum hypothesis).

Could it be that the answers to the problems depend on the choice of axioms—or even that they're independent of the traditional axioms (in the sense of Gödel's incompleteness theorem)? Historical experience in mathematics makes this seem extremely unlikely, because, to date, essentially all "natural" problems in mathematics seem to have turned out to be decidable in the (sometimes rather implicit) axiom system that's used in doing the mathematics.

In the computational universe, though—freed from the bounds of historical math tradition—it's vastly more common to run into undecidability. And, actually, my guess is that a fair fraction of long-unsolved problems even in traditional mathematics will also turn out to be undecidable. So that definitely raises the possibility that the problems here could be independent of at least some standard axiom systems.

OK, but assume there's no undecidability around, and one's not dealing with the few cases in which one can just answer a problem by saying "look at this explicitly constructed thing". Well, then to answer the problem, we're going to have to give a proof.

In essence what drives the need for proof is the presence of something infinite. We want to know something for any $n$, even infinitely large, etc. And the only way to handle this is then to represent things symbolically ("the symbol Infinity means infinity", etc.), and apply formal rules to everything, defined by the axioms in the underlying axiom system one's assuming.

In the best case, one might be able to just explicitly exhibit that series of rule applications—in such a way that a computer can immediately verify that they're correct. Perhaps the series of rule applications could be found by automated theorem proving (as in FindEquationalProof). More likely, it might be constructed using a proof assistant system.

It would certainly be exciting to have a fully formalized proof of the answer to any of the problems. But my guess is that it'll be vastly easier to construct a standard proof of the kind human mathematicians traditionally do. What is such a proof? Well, it's basically an argument that will convince other humans that a result is correct.

There isn't really a precise definition of that. In our step-by-step solutions in Wolfram|Alpha, we're effectively proving results (say in calculus) in such a way that students can follow them. In an academic math journal, one's giving proofs that successfully get past the peer review process for the journal.

My own guess would be that if one were to try to formalize essentially any nontrivial proof in the math literature, one would find little corners that require new results, though usually ones that wouldn't be too hard to get.

How can we handle this in practice for our prizes? In essence, we have to define a computational contract for what constitutes success, and when prize money should be paid out. For a constructive proof, we can get Wolfram Language code that can explicitly be run on any sufficiently large computer to establish the result. For formalized proofs, we can get Wolfram Language code that can run through the proof, validating each step.

But what about for a "human proof"? Ultimately we have no choice but to rely on some kind of human review process. We can ask multiple people to verify the proof. We could have some blockchain-inspired scheme where people "stake" the correctness of the proof, then if one eventually gets consensus (whatever this means) one pays out to people some of the prize money, in proportion to their stake. But whatever is done, it's going to be an imperfect, "societal" result—like almost all of the pure mathematics that's so far been done in the world.

### What Will It Take?

OK, so for people interested in working on the Problems, what skills are relevant? I don't really know. It could be discrete and combinatorial mathematics. It could be number theory, if there's a correspondence with number-based systems found. It could be some branch of algebraic mathematics, if there's a correspondence with algebraic systems found. It could be dynamical systems theory. It could be something closer to mathematical logic or theoretical computer science, like the theory of term rewriting systems.

Of course, it could be that no existing towers of knowledge—say in branches of mathematics—will be relevant to the problems, and that to solve them will require building "from the ground up". And indeed that's effectively what ended up happening in the solution for my 2, 3 Turing Machine Prize in 2007.

I'm a great believer in the power of computer experiments—and of course it's on the basis of computer experiments that I've formulated the Rule 30 Prize Problems. But there are definitely more computer experiments that could be done. So far we know a billion elements in the center column sequence. And so far the sequence doesn't seem to show any deviation from randomness (at least based on tests I've tried). But maybe at a trillion elements (which should be well within range of current computer systems) or a quadrillion elements, or more, it eventually will—and it's definitely worth doing the computations to check.

The direct way to compute *n* elements in the center column is to run rule 30 for *n* steps, using at an intermediate stage up to *n* cells of memory. The actual computation is quite well optimized in the Wolfram Language. Running on my desktop computer, it takes less than 0.4 seconds to compute 100,000 elements:

In[ ]:= CellularAutomaton[30, {{1}, 0}, {100 000, {{0}}}]; // Timing

Out[ ]= {0.383683, Null}

Internally, this is using the fact that rule 30 can be expressed as Xor[*p*, Or[*q*, *r*]], and implemented using bitwise operations on whole words of data at a time. Using explicit bitwise operations on long integers takes about twice as long as the built-in CellularAutomaton function:

In[ ]:= Module[{a = 1}, Table[BitGet[a, a = BitXor[a, BitOr[2 a, 4 a]]; i - 1], {i, 100 000}]]; // Timing

Out[ ]= {0.737693, Null}

But these results are from single CPU processors. It's perfectly possible to imagine parallelizing across many CPUs, or using GPUs. One might imagine that one could speed up the computation by effectively caching the results of many steps in rule 30 evolution, but the fact that across the rows of the rule 30 pattern all blocks appear to occur with at least roughly equal frequency makes it seem as though this would not lead to significant speedup.

Solving some types of math-like problems seem pretty certain to require deep knowledge of high-level existing mathematics. For example, it seems quite unlikely that there can be an "elementary" proof of Fermat's last theorem, or even of the four-color theorem. But for the Rule 30 Prize Problems it's not clear to me. Each of them might need sophisticated existing mathematics, or they might not. They might be accessible only to people professionally trained in mathematics, or they might be solvable by clever "programming-style" or "puzzle-style" work, without sophisticated mathematics.

## Generalizations and Relations

Sometimes the best way to solve a specific problem is first to solve a related problem—often a more general one—and then come back to the specific problem. And there are certainly many problems related to the Rule 30 Prize Problems that one can consider.

For example, instead of looking at the vertical column of cells at the center of the rule 30 pattern, one could look at a column of cells in a different direction. At 45°, it's easy to see that any sequence must be periodic. On the left the periods increase very slowly; on the right they increase rapidly. But what about other angles?

Or what about looking at rows of cells in the pattern? Do all possible blocks occur? How many steps is it before any given block appears? The empirical evidence doesn't see any deviation from blocks occurring at random, but obviously, for example, successive rows are highly correlated.

What about different initial conditions? There are many dynamical systems–style results about the behavior of rule 30 starting with equal probability from all possible infinite initial conditions. In this case, for example, it's easy to show that all possible blocks occur with equal frequency, both at a given row, and in a given vertical column. Things get more complicated if one asks for initial conditions that correspond, for example, to all possible sequences generated by a given finite state machine, and one could imagine that from a sequence of results about different sets of possible initial conditions, one would eventually be able to say something about the case of the single black cell initial condition.

Another straightforward generalization is just to look not at a single black cell initial condition, but at other "special" initial conditions. An infinite periodic initial condition will always give periodic behavior (that's the same as one gets in a finite-size region with periodic boundary conditions). But one can, for example, study what happens if one puts a "single defect" in the periodic pattern:

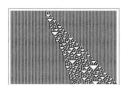 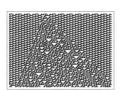 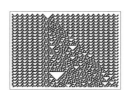 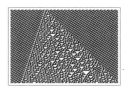

One can also ask what happens when one has not just a single black cell, but some longer sequence in the initial conditions. How does the center column change with different initial sequences? Are there finite initial sequences that lead to "simpler" center columns?

Or are there infinite initial conditions generated by other computational systems (say substitution systems) that aren't periodic, but still give somehow simple rule 30 patterns?

Then one can imagine going "beyond" rule 30. What happens if one adds longer-range "exceptions" to the rules? When do extensions to rule 30 show behavior that can be analyzed in one way or another? And can one then see the effect of removing the "exceptions" in the rule?

Of course, one can consider rules quite different from rule 30 as well—and perhaps hope to develop intuition or methods relevant to rule 30 by looking at other rules. Even among the 256 two-color nearest-neighbor rules, there are others that show complex behavior starting from a simple initial condition:

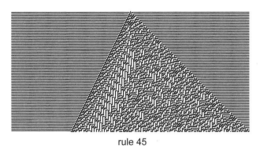 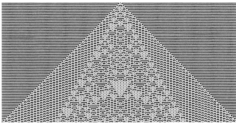

rule 45    rule 73

And if one looks at larger numbers of colors and larger neighbors one can find an infinite number of examples. There's all sorts of behavior that one sees. And, for example, given any particular sequence, one can search for rules that will generate it as their center column. One can also try to classify the center-column sequences that one sees, perhaps identifying a general class "like rule 30" about which global statements can be made.

But let's discuss the specific Rule 30 Prize Problems. To investigate the possibility of periodicity in rule 30 (as in Problem 1), one could study lots of different rules, looking for examples with very long periods, or very long transients—and try to use these to develop an intuition for how and when these can occur.

To investigate the equal-frequency phenomenon of Problem 2, one can look at different statistical features, and see both in rule 30 and across different rules when it's possible to detect regularity.

For Problem 3, one can start looking at different levels of computational effort. Can one find the $n^{th}$ value with computational effort $O(n^\gamma)$ for any $\gamma<2$ (I don't know any method to achieve this)? Can one show that one can't find the $n^{th}$ value with less than $O(\log(n))$ computational effort? What about with less than $O(\log(n))$ available memory? What about for different rules? Periodic and nested patterns are easy to compute quickly. But what other examples can one find?

As I've mentioned, a big achievement would be to show computation universality for rule 30. But even if one can't do it for rule 30, finding additional examples (beyond, for example, rule 110) will help build intuition about what might be going on in rule 30.

Then there's NP-completeness. Is there a way of setting up some question about the behavior of rule 30 for some family of initial conditions where it's possible to prove that the question is NP-complete? If this worked, it would be an exciting result for cryptography. And perhaps, again, one can build up intuition by looking at other rules, even ones that are more "purposefully constructed" than rule 30.

## How Hard Are the Problems?

When I set up my 2, 3 Turing Machine Prize in 2007 I didn't know if it'd be solved in a month, a year, a decade, a century, or more. As it turned out, it was actually solved in about four months. So what will happen with the Rule 30 Prize Problems? I don't know. After nearly 40 years, I'd be surprised if any of them could now be solved in a month (but it'd be really exciting if that happened!). And of course some superficially similar problems (like features of the digits of $\pi$) have been out there for well over a century.

It's not clear whether there's any sophisticated math (or computer science) that exists today that will be helpful in solving the problems. But I'm confident that whatever is built to solve them will provide structure that will be important for solving other problems about the computational universe. And the longer it takes (think Fermat's last theorem), the larger the amount of useful structure is likely to be built on the way to a solution.

I don't know if solutions to the problems will be "obviously correct" (it'll help if they're constructive, or presented in computable form), or whether there'll be a long period of verification to go through. I don't know if proofs will be comparatively short, or outrageously long. I don't know if the solutions will depend on details of axiom systems ("assuming the continuum hypothesis", etc.), or if they'll be robust for any reasonable choices of axioms. I don't know if the three problems are somehow "comparably difficult"—or if one or two might be solved, with the others holding out for a very long time.

But what I am sure about is that solving any of the problems will be a significant achievement. I've picked the problems to be specific, definite and concrete. But the issues of randomness and computational irreducibility that they address are deep and general. And to know the solutions to these problems will provide important evidence and raw material for thinking about these issues wherever they occur.

Of course, having lived now with rule 30 and its implications for nearly 40 years, I will personally be thrilled to know for certain even a little more about its remarkable behavior.

# GALLERY OF ART

**Ryan Bell**
*Rule-110 (2020)*
Digital artwork

Anna Brzozowska
*Cellular Automata (2015)*
Ink on paper

**ciphrd (fxhash username)**
*RGB Elementary Cellular Automaton (2021)*
NFT, digital artwork

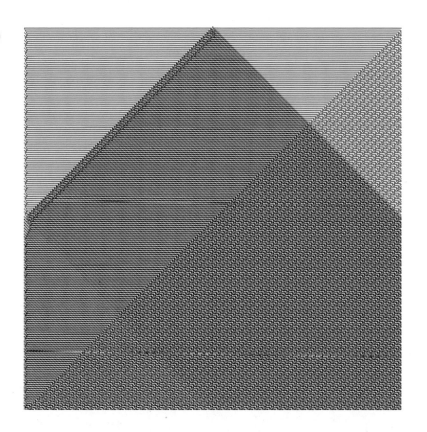

**x0y0z0 (fxhash username)**
*Dynamic generations (2021)*
NFT, digital artwork

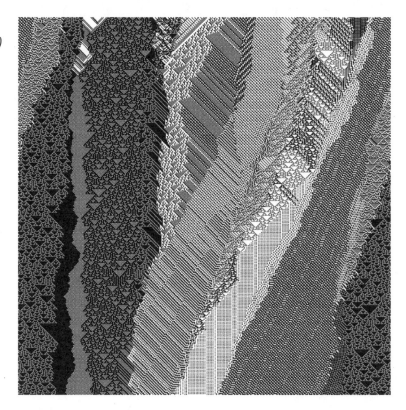

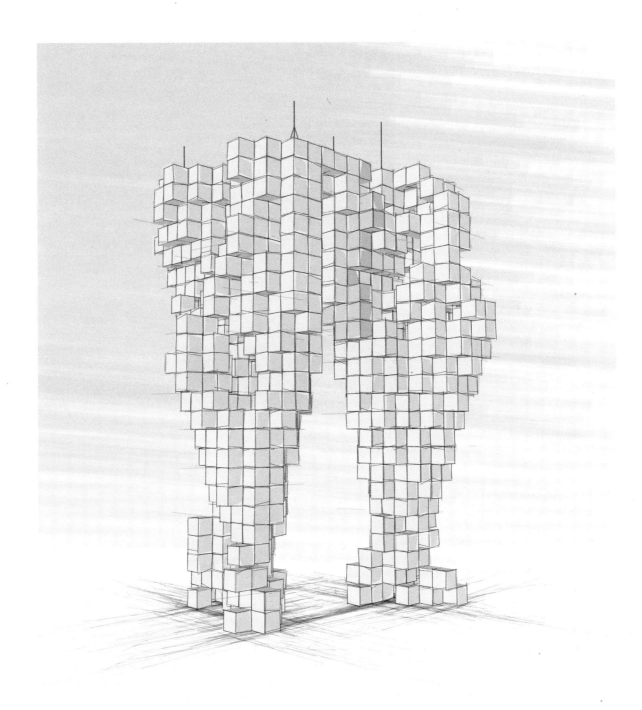

**@phil_osophie** (fxhash username)
*Cellular Skyscrapers (2022)*
NFT, digital artwork

**Igor Bakshee**
*Sunny Image of Rule 110 (1997)*
Digital artwork

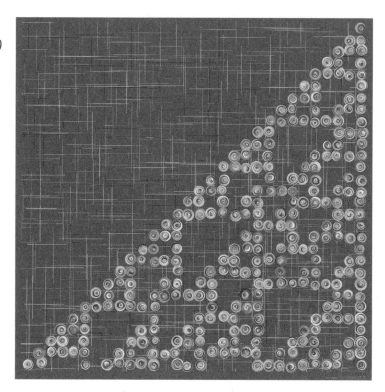

**Igor Bakshee**
*Cellular Automata Stones (1997)*
Digital artwork

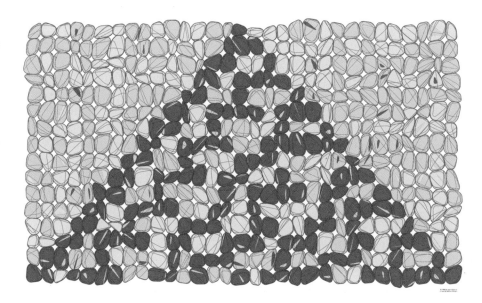

**Trevor Scott and Alyssa Scott**
*Digital Mortal Logo (2021)*
Digital artwork

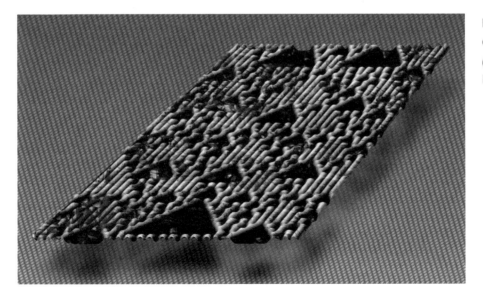

**Martin Zillmann**
*OneAngstromCircuit (2021)*
Digital artwork

**Ryan Bell**
*Rule-73 (2020)*
Digital artwork

**Phil Wheeler**
*Pixel Fantasy (2017)*
Digital artwork

**Thomas Lowe**
*Crystal Rock (2012)*
Digital artwork

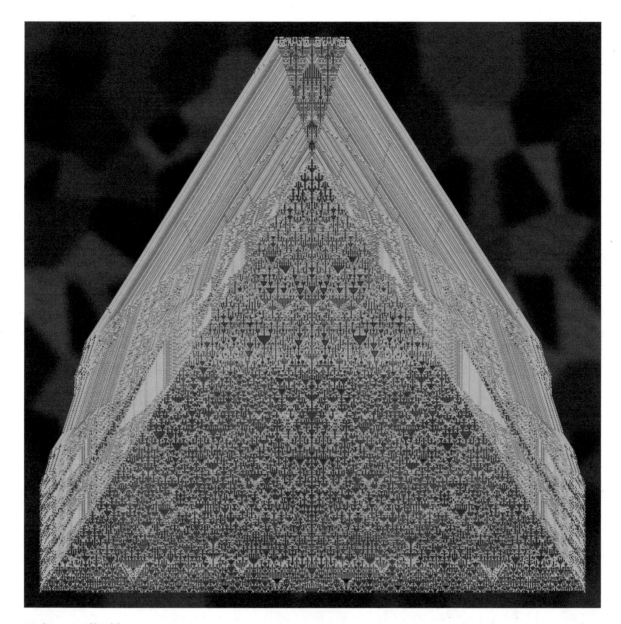

**Andrew Kotlinski**
*Intagonies (2016)*
Digital artwork

**Andrew Kotlinski**
*Watergasms (2017)*
Digital artwork

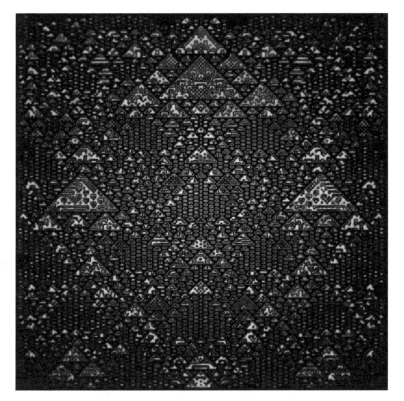

**Martin Zillmann**
*TrianglZ (2021)*
Digital artwork

**Damir Bogdan**
*Shamans Journey 220 (2014)*
Digital artwork

**Rafael Lanfranco**
*La retromáquina universal de*
*computación constante (2021)*
Mixed technique, drawing and digital art

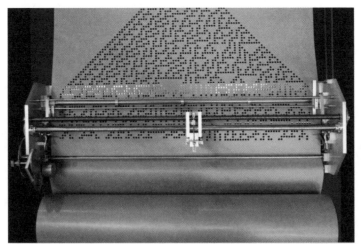
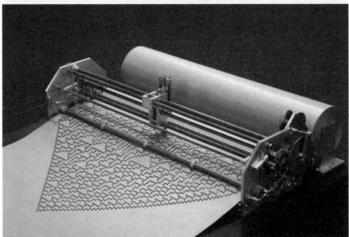
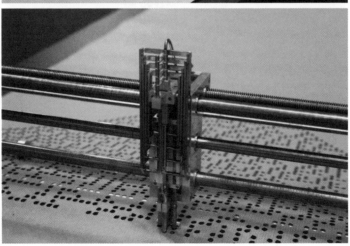

**Kristoffer Myskja**
*Rule 30 (2008)*
Kinetic sculpture

**Owen Schuh**
*Phase Change (2010)*
Oil on glass Petri dish

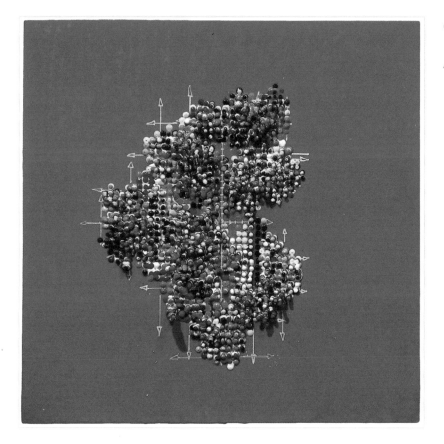

**Owen Schuh**
*The Fall (Rule #23/3) (2007)*
Acrylic and enamel on panel

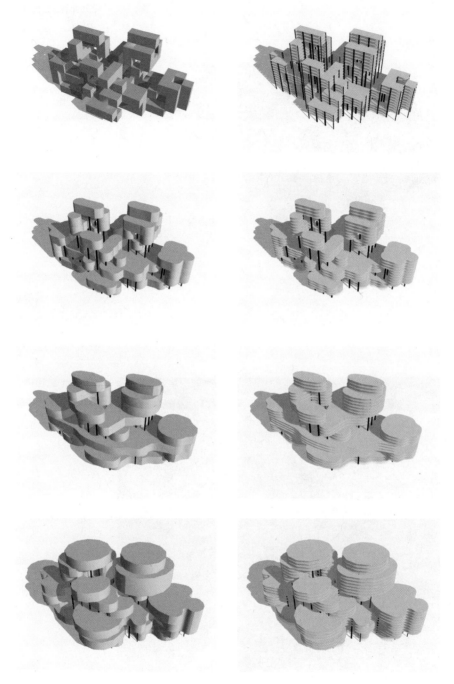

**Robert Krawczyk**
*Architectural Interpretation of Cellular Automata (2002)*
Generative art

**Jeff Cook**
*Wolfrule Art (2011)*
Wood

 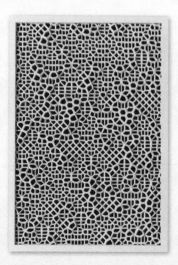 

**Ori Adiri**
*Parametric Cellular Automata (2019)*
3D printed with sandstone

*NKS display at Cooper Hewitt, Smithsonian Design Museum (2004)*

**Maciej Zawidzki**
*Dynamic shading of a building envelope controlled by one-dimensional CA (2015)*
Building prototype

Antonio Carlos Barbosa de Oliveira
*Rule 30 design in swimming pool (2018)*

**Quintin Doyle**
*Rule 30 Cambridge North railway station (2017)*

**Madeleine Shepherd**
*Rule 30 headbands (2017)*
Lambswool

**Cam Fox**
*Rule 105 quilt (2013)*
Quilting cotton

**René Sultra and Maria Barthélémy in collaboration with François Roussel, Nazim Fatès and Moulin Gau**
*Sentimental journey (2008)*
Polyester jacquard weaving

**Madeleine Shepherd**
*Rule 30 shoulder bag (2019)*
Wool, cotton polyester blend

**Beam Contrechoc**
*Protest Knit (2012)*
Knitting yarn

**Tomáš Miškov**
*Rule 30 T-shirt (2020)*
Cotton and polyester

**Cam Fox**
*Rule 109 tea cozy (2013)*
100% acrylic worsted weight yarn

**Cam Fox**
*Rule 110 tea cozy (2008)*
100% acrylic worsted weight yarn

**Lesley Starke**
*Chaos: A new kind of scarf (2021)*
Knitted wool

**Fabienne Haas**
*Rule 73 scarf (2011)*
Knitting yarn

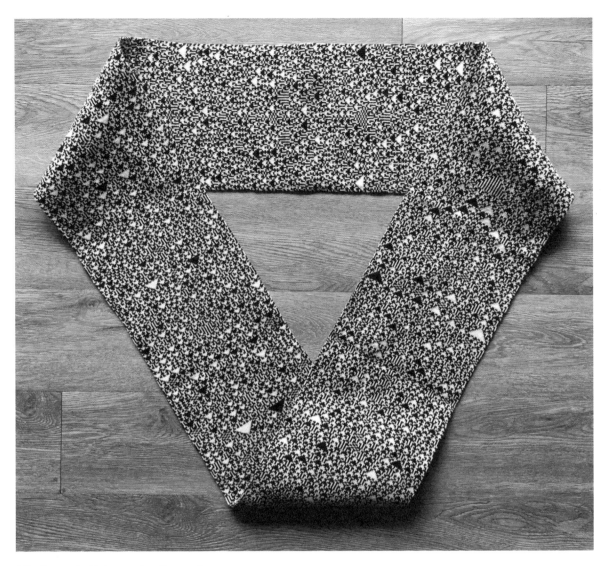

Elisabetta A. Matsumoto, Henry Segerman
and Fabienne Haas
*Knitted Möbius scarf rule 150 (2018)*
Merino wool

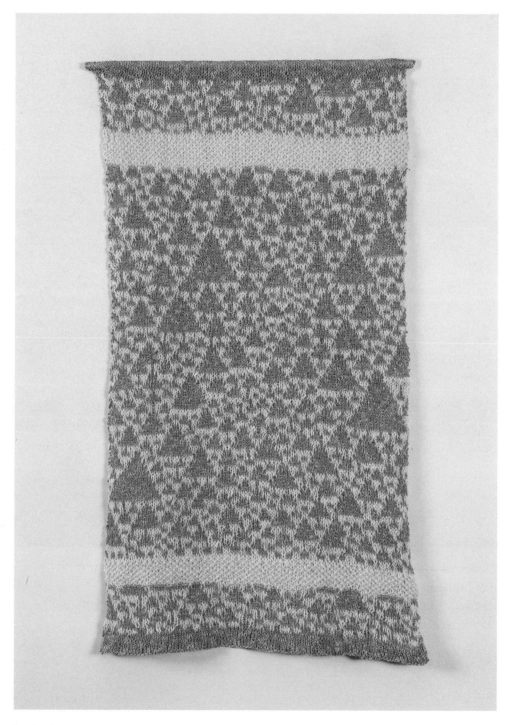

**Lesley Turner**
*Cellular Automata: Rule 30 (2011)*
Hemp, rayon, cotton, wool

**Tatsuki Hayama**
*The cover art of Generative Art with Math (2017)*
Fabric

**Marina Toeters and Loe Feijs**
*A Cellular Automaton for Pied-de-poule (2017)*
Woven polyester textile

**Genaro Rivas in collaboration with Christian Pasquel and Omar Ortiz**
*#LaCasacaNACIONAL (2017)*
Recycled polyester

Owen Schuh
*Non-periodic Growth (soft) (2007)*
Red felt

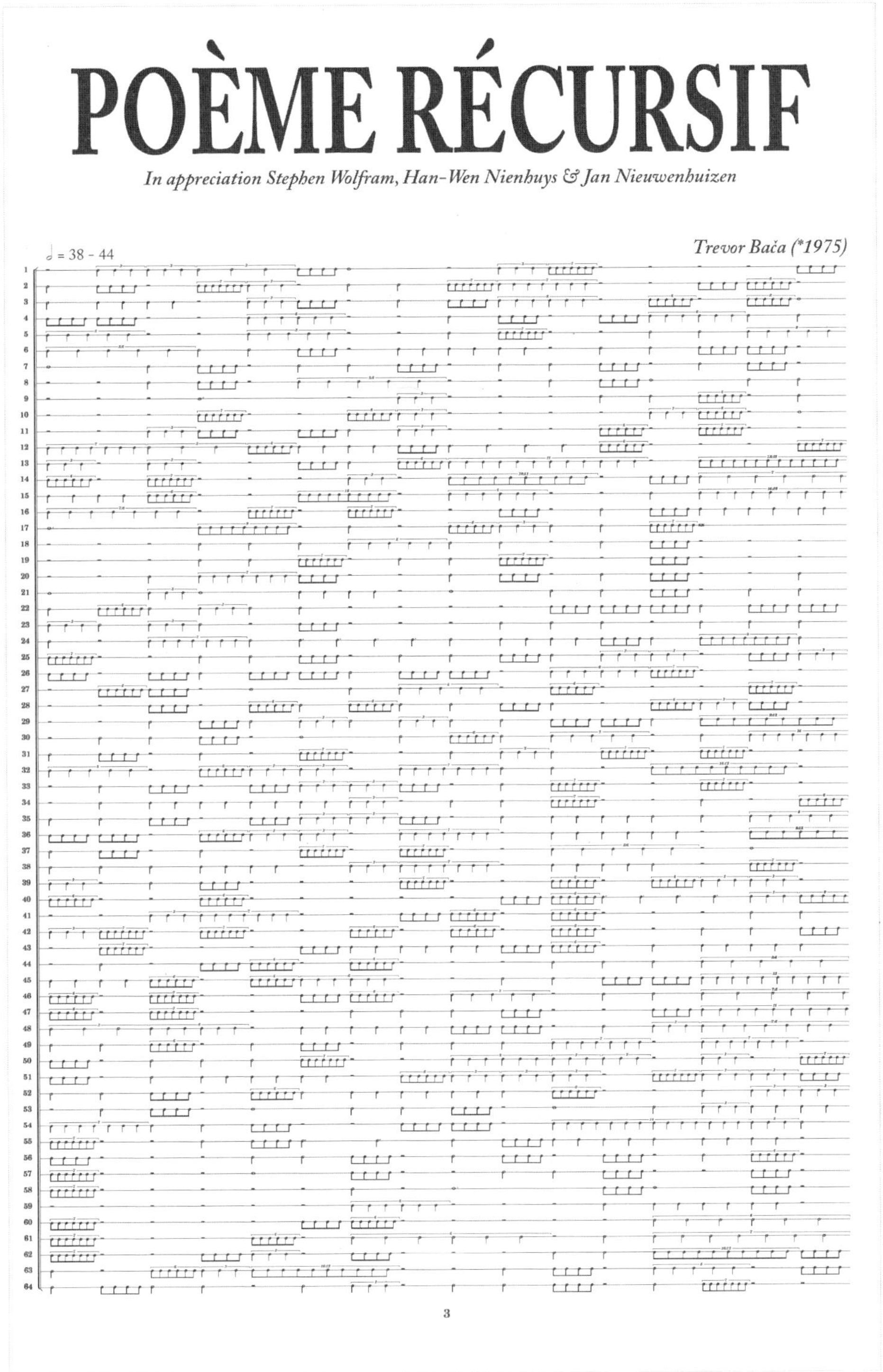

# POÈME RÉCURSIF

*In appreciation Stephen Wolfram, Han-Wen Nienhuys & Jan Nieuwenhuizen*

*Trevor Bača (*1975)*

**Trevor Bača**
*Poème récursif (2005)*
64 pieces of percussion
wolfr.am/nks-gallery-1

**Srinivas Mangipudi**
*Automata cotton fabrics (2022)*
NFT, digital video
wolfr.am/nks-gallery-2

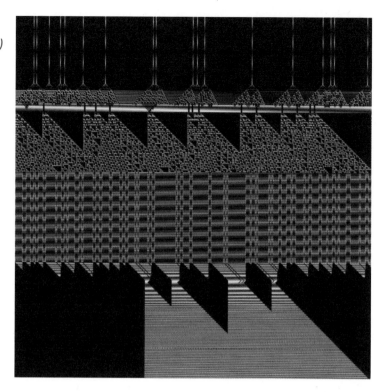

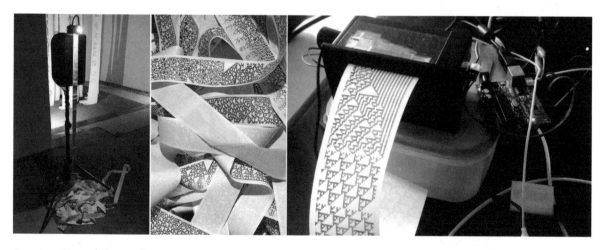

**Agoston Nagy (Binaura)**
*00011110 (2019)*
Thermal printer, Arduino, speaker
wolfr.am/nks-gallery-3

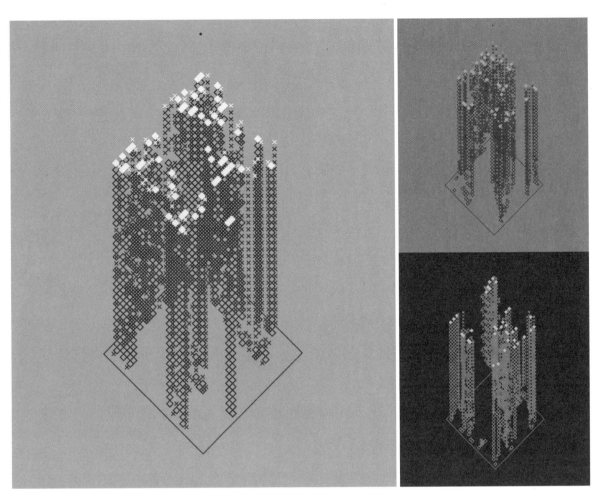

**Agoston Nagy (Binaura)**
*L-2D CA (2021)*
NFT, generative art
wolfr.am/nks-gallery-4

**Trevor Scott and Alyssa Scott**
*Voltaire Sealed, Cardano Mythos Collection (2021)*
NFT, digital video
wolfr.am/nks-gallery-5

**Dmitry Morozov**
*Conus (2013/2019)*
Multimedia installation
wolfr.am/nks-gallery-6

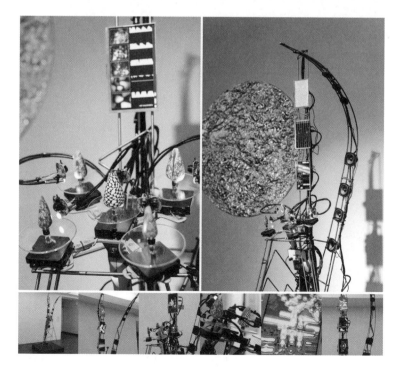

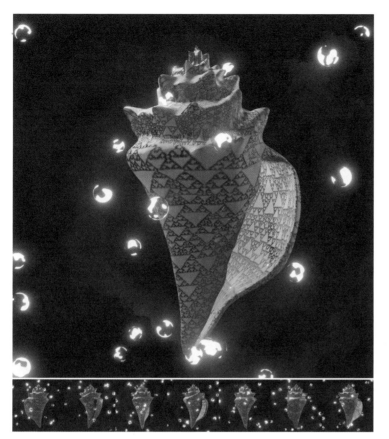

**Polina Ostapchuk**
*Shellural automata (2021)*
NFT, digital video
wolfr.am/nks-gallery-7

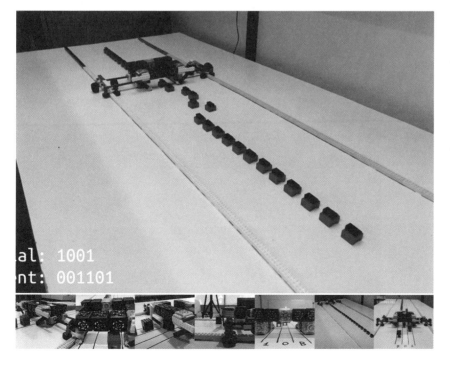

**Genaro Martínez, Ricardo Figueroa and Andrew Adamatzky**
*Robotic Tag System (Post machine) (2021)*
Modular robots (Cubelets) and LEGO pieces
wolfr.am/nks-gallery-8

**Stephen Wolfram**
*Postcards (1984)*

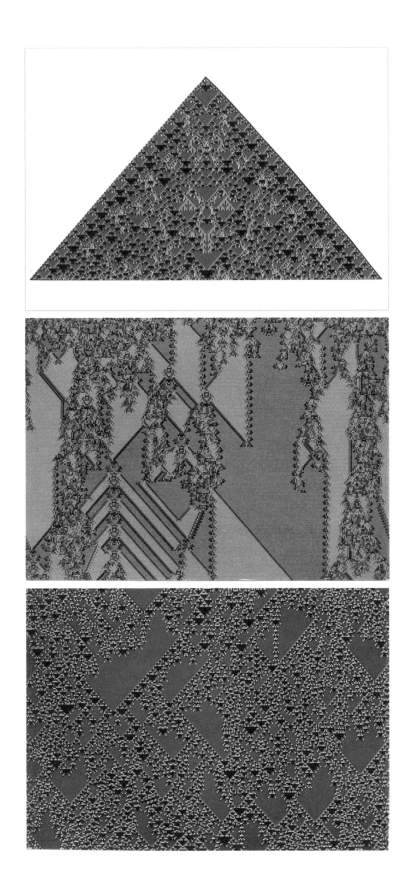

**Stephen Wolfram**
*Untitled (2004)*
Digital artwork

**Stephen Wolfram**
*Firing Squad (2004)*
Digital artwork

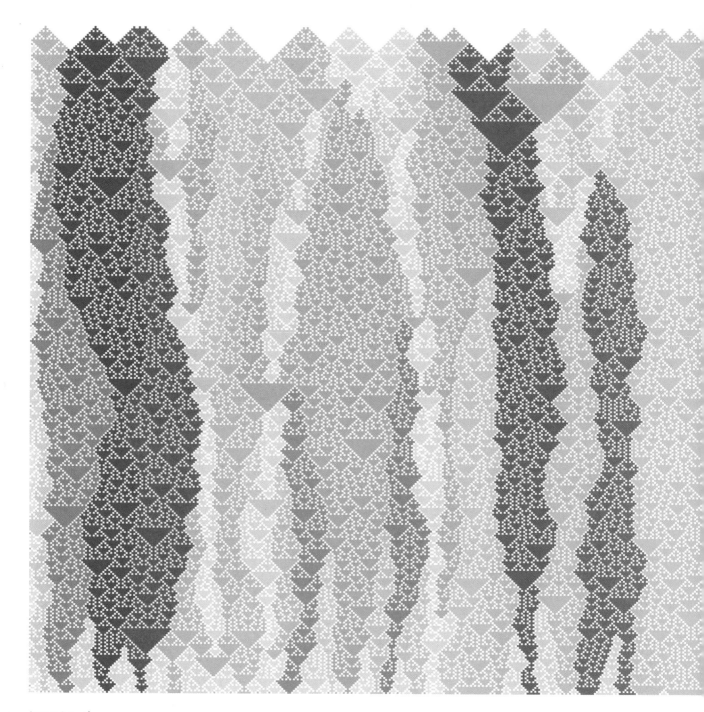

**Jorge Laval**
*Power Law Behavior in Elementary Cellular Automata (2021)*
Interactive Demonstration
wolfr.am/nks-gallery-9

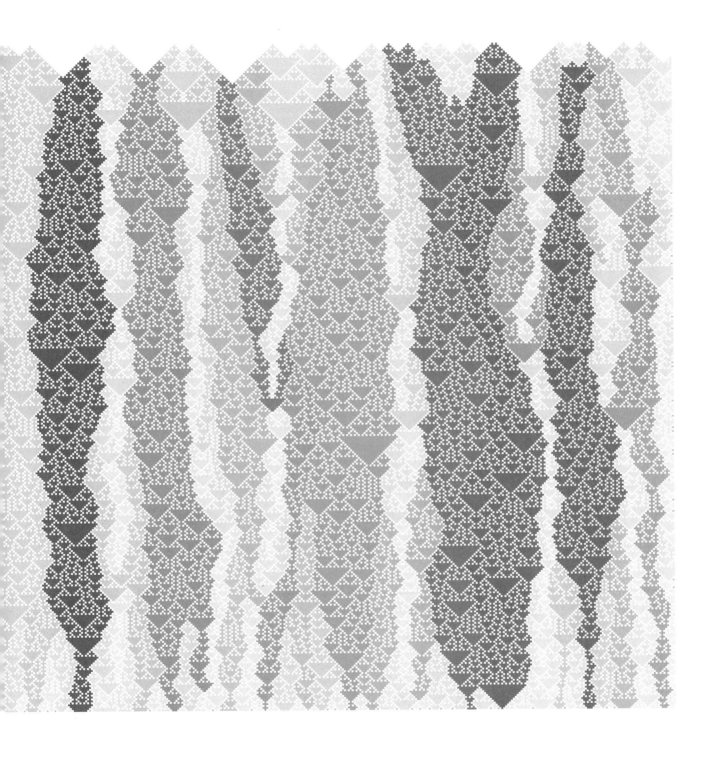

**Vitaliy Kaurov**
*Order, Chaos, and the Formation of a Cantor Set Attractor in Elementary Cellular Automata (2013)*
Interactive Demonstration
wolfr.am/nks-gallery-10

**Michael Schreiber**
*Outer-Totalist Hexagons Between Past and Future (2011)*
Interactive Demonstration
wolfr.am/nks-gallery-11

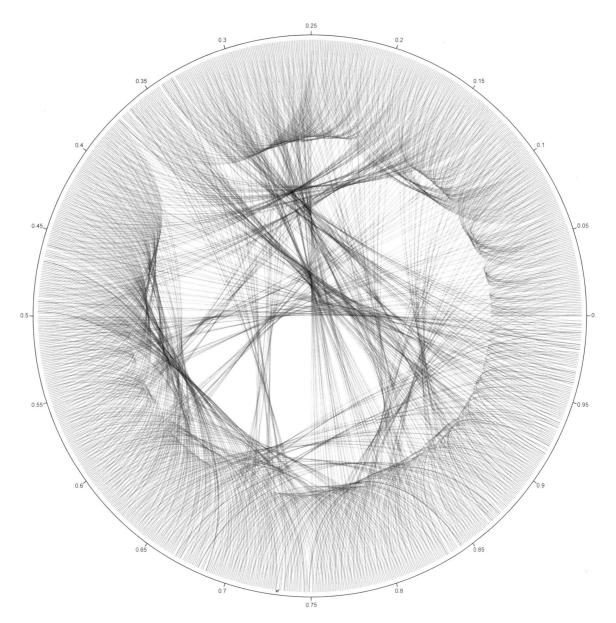

**Vitaliy Kaurov**
*Iconography for Elementary Cellular Automata Based on Radial Convergence Diagrams (2013)*
Interactive Demonstration
wolfr.am/nks-gallery-12

**Jeremy Davis**
*Rule 30 Figure Eight (2021)*
Digital artwork

**Jeremy Davis**
*Rule 30 Trefoil (2021)*
Digital artwork

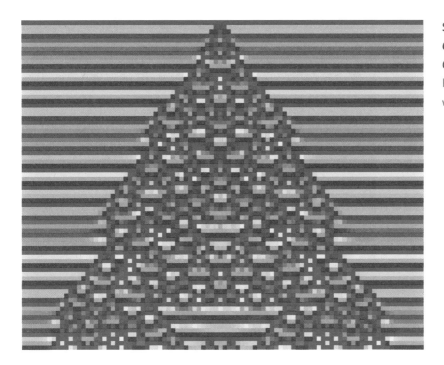

**Stephen Wolfram**
*Constant-Addition Continuous Cellular Automaton (2011)*
Interactive Demonstration
wolfr.am/nks-gallery-13

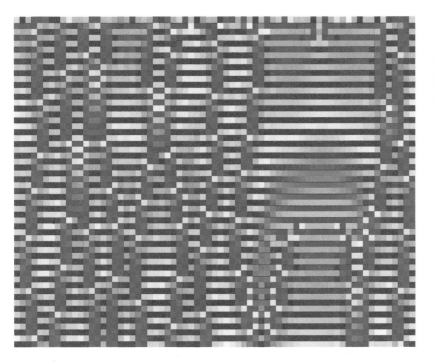

**Stephen Wolfram**
*Constant-Addition Continuous Cellular Automaton (2011)*
Interactive Demonstration
wolfr.am/nks-gallery-13

**Daniel de Souza Carvalho**
*Patterns from the Mean of Two-Color Totalistic 2D Cellular Automata (2011)*
Interactive Demonstration
wolfr.am/nks-gallery-14

**Daniel de Souza Carvalho**
*Patterns from the Mean of Two-Color Totalistic 2D Cellular Automata (2011)*
Wolfram Demonstration
wolfr.am/nks-gallery-14

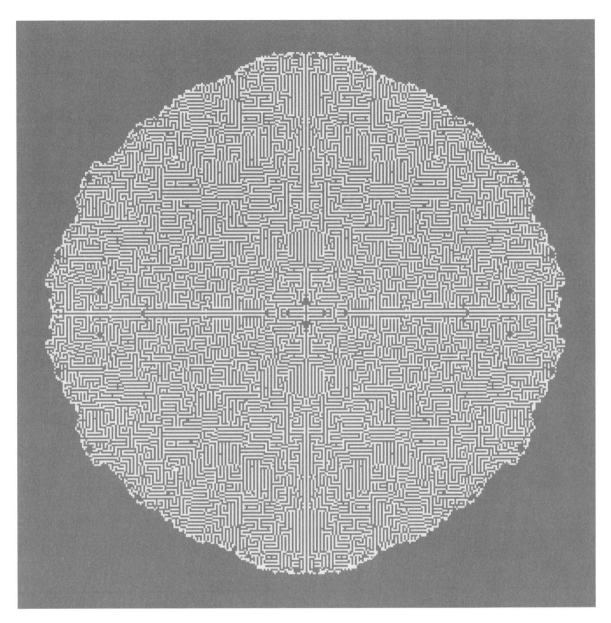

**Stephen Wolfram**
*Outer Totalistic Code 756 (2002)*
Digital artwork

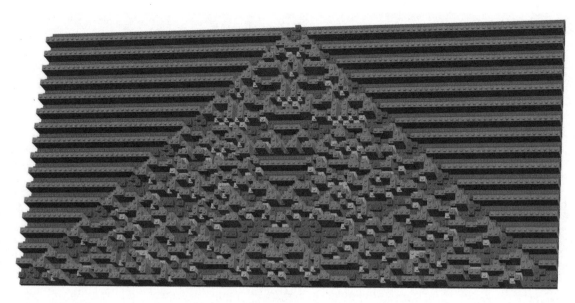

**Jeremy Davis**
*Totalistic LEGO Rule #1124126 (2022)*
Digital artwork

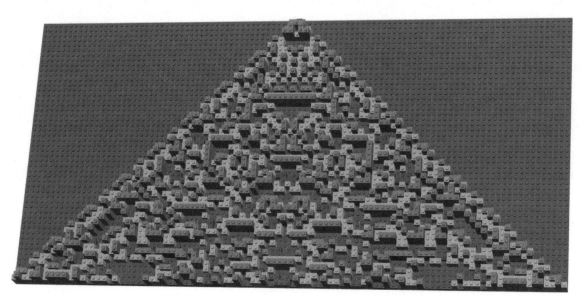

**Jeremy Davis**
*Totalistic LEGO Rule #1006540 (2022)*
Digital artwork

Stephen Wolfram and Heidi Kellner
*Sequences of cellular automata with 2, 3, 4, and 5 colors*
Digital artwork

**Wolfram Design Group**
*WolframTones Swatch Book (2021)*

**Wolfram Design Group**
*Stephen Wolfram Business Cards (2021)*

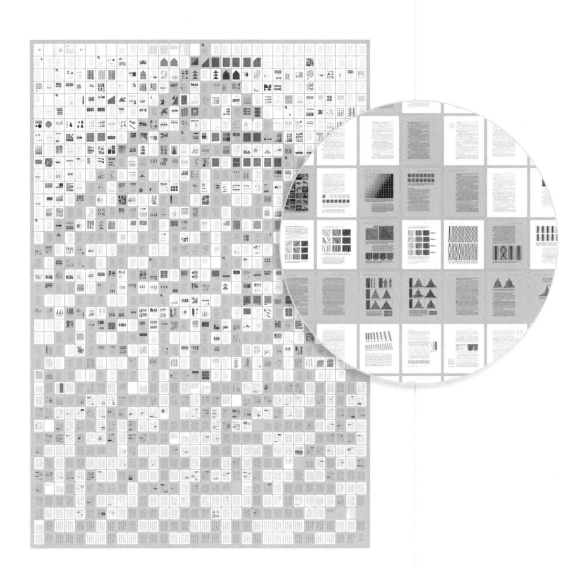

**Wolfram Design Group**
*A New Kind of Science Thumbnail Poster (2002)*

Circuit Seaweed

The Irreducible Carpet

Plant City

Crystal Queen

Stephen Wolfram
**NFT Collection**
*Cellular Automata from the Computational Universe (2021)*
Digital artwork

*The Tracks Emerge*

*Infinite Willow*

*Organic Power Tie*

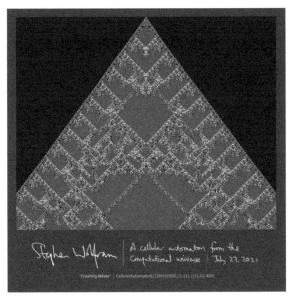

*Crashing Waves*

 Browse the full gallery of art online at
**wolframscience.com/gallery-of-art**